Viva la Difference in Statistics

Third Edition

Kenneth A. Weaver

Emporia State University

Kendall Hunt

publishing company

Cover image © Shutterstock, Inc.

Kendall Hunt
publishing company

www.kendallhunt.com
Send all inquiries to:
4050 Westmark Drive
Dubuque, IA 52004-1840

Copyright © 1998, 2003, 2013 by Kenneth A. Weaver

ISBN 978-1-4652-8260-6

Kendall Hunt Publishing Company has the exclusive rights to reproduce this work,
to prepare derivative works from this work, to publicly distribute this work,
to publicly perform this work and to publicly display this work.

All rights reserved. No part of this publication may be reproduced,
stored in a retrieval system, or transmitted, in any form or by any
means, electronic, mechanical, photocopying, recording, or otherwise,
without the prior written permission of the copyright owner.

Printed in the United States of America

Contents

Preface to the Third Edition

The goal of the third edition remains as always to assist students in having a positive, meaningful experience while learning basic statistics. I thank the hundreds of students who have used the book with deep appreciation for helping me be a better teacher of statistics and for their numerous suggestions which have been incorporated into all three editions of the book. The book's features are:

- The symbols and formulas typically used in the course are presented immediately after the Preface for easy reference.

- Chapter 1 is a review of arithmetic and algebra, concluding with a diagnostic test to assist students in strengthening skills at the beginning of the course.

- Numerous examples of the different statistical procedures appear throughout the book.

- Added to each of the inferential statistics chapters is a "Researcher Problem" where students are the researchers analyzing the data and interpreting the results in the context of a hypothesis or question.

- The Researcher Problems for Chapters 8, 9, 10, and 11 also include the data set in case the student wants (or professor requires) to analyze using statistical software such as SPSS or SAS.

- Additional exercises and sample problems are available both in and at the end of each chapter and they cover disciplines including psychology and nursing. Sample problems assess content knowledge presented earlier in the book to ascertain cumulative understanding.

- Several figures are included to illustrate the statistical content including a graph of weights and lengths for girls from 0 to 3 years of age from the Centers for Disease Control in Chapter 4, a random number table in Chapter 6, and a table of Type I and Type II errors in Chapter 9.

- A positive "can do" tone reflects my approach that statistics is understandable and useful.

- *Critical Thinking Questions* and their answers increase convenience and efficient use of study time.

- The answers plus explanations to all end-of-the-chapter problems are located at the end of the book so you can check all of your work.

I thank my students whose influence permeates and strengthens this book. With much love and appreciation, I thank my wife Katherine and my children Merriam, Andrew, Katherine, and Janet for their love and support throughout my career.

I thank Dr. Jane Monroe, my statistics professor at Teachers College, Columbia University, whose clarity in teaching I humbly strive to model.

Finally, I thank Dr. Loren Tompkins, Emporia State University Professor Emeritus, whom I respect and admire for his knowledge of statistics, his unwavering commitment to students across the decades, his review of the first edition of this book, and his friendship. To him this book is dedicated.

All proceeds from the sale of this book go to student scholarships.

Symbols

raw score = X

sum of the raw scores = ΣX

sample mean = M (in some textbooks the sample mean is symbolized as $\overline{X}$)

number of participants in the population or sample = N

number of participants in a treatment group or condition = n

deviation score = $X - M = x$

the raw scores squared then summed = ΣX^2

the raw scores summed then squared = $(\Sigma X)^2$

sum of squares = $SS = \Sigma(X - M)^2 = \Sigma x^2 = \Sigma X^2 - \dfrac{(\Sigma X)^2}{N}$

variance = V

sample standard deviation (not estimating population standard deviation) = SD

location of a raw score in a sample = z

location of a raw score in a population = z

probability = p

population mean **AND** the mean of the sampling distribution of means = μ

population standard deviation = σ

standard error of the mean = σ_M

location of a sample mean on the sampling distribution of means (parameters known) = z

null hypothesis = H_0

alternative hypothesis = H_1 or H_a

alpha or the probability of making a Type I error (i.e., rejecting H_0 when H_0 is true) = α

estimated standard deviation of the population or sample standard deviation = s

estimated standard error of the mean = SE_M

degrees of freedom $= df$

location of the sample mean on the sampling distribution of means (parameter(s) unknown) $= t$

beta or the probability of making a Type II error (i.e., not rejecting H_0 when it is false) $= \beta$

estimated standard error of difference $= SE_D$

location of the difference between two sample means on the sampling distribution of differences (parameters unknown) $= t$

sum of squares $_{\text{between groups}}$ $= SS_b$

sum of squares $_{\text{within groups}}$ $= SS_w$

sum of squares $_{\text{total}}$ $= SS_t$

number of treatment condition means $= k$

variance $_{\text{between groups}}$ $= V_b$ $=$ mean square $_{\text{between groups}}$ $= MS_b$

variance $_{\text{within groups}}$ $= V_w$ $=$ mean square $_{\text{within groups}}$ $= MS_w$

location of the mean square $_{\text{between groups}}$ to the mean square $_{\text{within groups}}$ ratio on the sampling distribution of ratios of variances $= F$

sum of the cross products $= \Sigma XY$

correlation between two measurements in the population $= \rho$

correlation between two measurements in the sample $= r$

coefficient of determination $= r^2$

predicted value of $Y = \hat{Y}$

observed frequency $= f_o$

expected frequency $= f_e$

chi square $= X^2$

Formulas

Chapter 3

1. $\Sigma X = X_1 + X_2 + \ldots + X_N$

2. sample mean $= M = \dfrac{\Sigma X}{N}$

3. population mean $= \mu = \dfrac{\Sigma X_{population}}{N_{population}}$

4. inclusive range = highest score - lowest score + 1

5. exclusive range = highest score - lowest score

Chapter 4

1. deviation score $= X - M$

2. sum of squares $_{deviational} = SS = \Sigma \left(X - M \right)^2 = \Sigma x^2$

3. $\Sigma X^2 = X_1^2 + X_2^2 + \ldots + X_N^2$

4. sum of squares $_{computational} = SS = \Sigma X^2 - \dfrac{\left(\Sigma X \right)^2}{N}$

5. variance $= V = \dfrac{SS}{N} = \dfrac{\Sigma \left(X - M \right)^2}{N} = \dfrac{\Sigma x^2}{N} = \dfrac{\Sigma X^2 - \dfrac{\left(\Sigma X \right)^2}{N}}{N}$

6. standard deviation $= SD = \sqrt{\dfrac{SS}{N}} = \sqrt{\dfrac{\Sigma \left(X - M \right)^2}{N}} = \sqrt{\dfrac{\Sigma x^2}{N}} = \sqrt{\dfrac{\Sigma X^2 - \dfrac{\left(\Sigma X \right)^2}{N}}{N}}$

Chapter 5

1. $z = \dfrac{X - M}{SD}$

2. $X = \left(z \right)\left(SD \right) + M$

Chapter 7

1. sampling error $= M - \mu$

2. $\mu_M = \mu = \dfrac{\Sigma M}{N}$

3. $\sigma_M = \dfrac{\sigma}{\sqrt{N}}$

4. $z = \dfrac{M - \mu}{\sigma_M}$ (z test)

Chapter 8

1. $s = \sqrt{\dfrac{SS}{N-1}}$

2. $SE_M = \dfrac{s}{\sqrt{N}}$

3. $t = \dfrac{M - \mu}{SE_M}$ (One-sample t test)

4. $df = N - 1$

Chapter 9

1. $SE_D = \sqrt{SE_{M_1}^2 + SE_{M_2}^2}$ (separate variance estimate)

2. $SE_D = \sqrt{\dfrac{(N_1 - 1)s_1^2 + (N_2 - 1)s_2^2}{N_1 + N_2 - 2}}$ (pooled variance estimate)

3. $t = \dfrac{M_1 - M_2 - 0}{SE_D} = \dfrac{M_1 - M_2}{SE_D}$ (two-sample t test)

4. $df = (N_1 - 1) + (N_2 - 1) = N_1 + N_2 - 2$

5. effect size $= d = t\sqrt{\dfrac{n_1 + n_2}{(n_1)(n_2)}}$

Chapter 10

1. $F = \dfrac{V_b}{V_w} = \dfrac{MS_b}{MS_w} = \dfrac{\dfrac{SS_b}{df}}{\dfrac{SS_w}{df}} = \dfrac{\dfrac{(M_1 - M_t)^2 n_1 + (M_2 - M_t)^2 n_2 + (M_3 - M_t)^2 n_3}{k-1}}{\dfrac{\Sigma(X_1 - M_1)^2 + \Sigma(X_2 - M_2)^2 + \Sigma(X_3 - M_3)^2}{N-k}}$

2. $SS_{total} = SS_b + SS_w$

3. effect size $= \eta^2 = \dfrac{SS_b}{SS_t}$

Chapter 11

1. $\Sigma XY = X_1 Y_1 + X_2 Y_2 + \ldots + X_N Y_N$

2. $r = \dfrac{\dfrac{\Sigma XY - (M_X)(M_Y)(N)}{N-1}}{s_X s_Y} = \dfrac{\Sigma(X - M_X)(Y - M_Y)}{\sqrt{\Sigma(X - M_X)^2 \Sigma(Y - M_Y)^2}}$ or $\dfrac{\Sigma(X - M_X)(Y - M_Y)}{\sqrt{SS_X \cdot SS_Y}}$

3. $df = N_{\text{pairs of scores}} - 2$

4. $\hat{Y} = a + bX$

5. $SS_Y = $ the sum of squares of $Y = \Sigma(Y - M_Y)^2$

6. $SS_{Y \text{ not explained by } X} = SS_{\text{errors of prediction}} = \Sigma(Y - \hat{Y})^2$

7. $SS_{Y \text{ explained by } X} = \Sigma(\hat{Y} - M_Y)^2$

8. $SS_Y = \Sigma(Y - M_Y)^2 = \Sigma(Y - \hat{Y})^2 + \Sigma(\hat{Y} - M_Y)^2$

9. $r^2 = $ coefficient of determination $= \dfrac{SS_{Y \text{ explained by } X}}{SS_Y} = \dfrac{\Sigma(\hat{Y} - M_Y)^2}{\Sigma(Y - M_Y)^2} =$

the proportion of the sum of squares of Y explained by X

10. $SE_{\text{est}} = \sqrt{\dfrac{\Sigma(Y - \hat{Y})^2}{N-2}} = s_Y \sqrt{\dfrac{N(1-r^2)}{N-2}}$

Chapter 12

1. $\chi^2 = \Sigma \dfrac{(f_o - f_e)^2}{f_e}$

Chapter 1

Introduction and Review of Arithmetic and Algebra

Welcome to the wonderful world of statistics! You are starting an adventure during which you will become familiar with a tool of great value. The statistics course is a rarity (and rareness is a statistical term). It is one of the few courses where the introductory student actually learns skills being currently used by experts in the field. The graduating psychology major or nurse or physical therapist or biologist usually possesses a foundation of competencies but not the advanced knowledge of experts in their disciplines. When it comes to statistics, however, a quick look at research articles in journals from a variety of disciplines shows *t* tests, analyses of variance, and correlations being regularly used, the same procedures you will learn in your course.

The *Viva La Difference* book has this title for four reasons. First, just about all of the formulas in the course require subtraction and thus finding a difference. Look at the pages of formulas which follow the Preface. *Subtraction dominates these formulas.* Subtraction is one of the central themes of the statistics course around which to build your understanding.

Second, although there are many mathematical formulas used in the course, and statistics is a branch of mathematics, a statistics course is different from the math courses you have had. Statistics includes the crucial understanding about how the results of those formulas describe or detect patterns in the data. Such information can be used to help researchers make decisions about the existence of the effectiveness of a treatment or a relationship.

Third, you will be different as a result of this course. The statistics course consistently imparts many valuable skills and much valuable knowledge. Students leave the course with a much clearer idea of the value of statistics. What once seemed like gibberish to you will make good sense. You may remember from your introductory psychology course the "just noticeable difference" which is the smallest difference between two stimuli that you can detect. In contrast to the just noticeable difference, your understanding of statistics after the course will far exceed what you now know.

Fourth, a knowledge of statistics will make a difference in your life. Your ability to critically think about the meaningfulness of data will enrich your understanding immeasurably. In addition, I challenge you to name one occupation which does not in some form or fashion utilize statistics, either formally or informally. There simply is not one. Let's extend the point—being a member of a civic organization, Parent–Teacher Organization, 4-H club, Boy or Girl Scout troop, or church often requires working with information of a statistical nature. The average weekly attendance, the number of people to plan on coming to the school's open house, and the accuracy of a projection of anticipated revenue for the next fiscal year are questions which statistics or statistical thinking answers.

Two Perspectives on Learning Statistics: Doer of Research and Consumer of Research

Your academic studies may require you to be a *doer of research*, where you collect data for your own research study or assist a faculty member or graduate student in their projects. The data you collect will need to be analyzed statistically to test your hypotheses or answer your research questions. Your statistics course will show you how to do this.

Your studies will also require you to be a *consumer of research*. Class assignments, projects, and professional development often mean heading to the library and reading research articles contained in research journals. Most of those articles will contain *Results* sections where the researchers report their statistics. Your statistics course will enable you to attain greater insight into the results of these articles and why the data were analyzed the way they were.

Regardless of whether you *do* research or *consume* research (and most likely you will do a bit of both), a background in statistics will train you to collect, analyze, and report statistical information accurately and responsibly. This background will also help you understand statistical information when you encounter it and ask insightful questions to clarify ambiguity commonly experienced when encountering statistical information in advertisements, newspapers, magazines, and nonscientific reports. The statistics course also provides you with an approach to decision–making, useful even in non–statistical situations.

Take every advantage of what could well be your only formal training in statistics for your entire life. This course will strengthen your thinking, reasoning, and computational skills and enrich your understanding of the scientific method. Besides an improved quality of life, your competence with the material in your statistics course may mean a job or advancement one day.

Math Anxiety

Math anxiety, math phobia, math avoidance, sadistics! The thought of statistics can trigger a variety of uncomfortable thoughts and emotions. The upbeat title of this book stands in stark contrast to books such as *Statistics for the Terrified* (Kranzler, 2011), *Statistics for Math Haters* (Lovejoy, 1975), and *Statistics Without Tears* (Rowntree, 1981), which may more accurately reflect what you are experiencing as you start your statistics course.

Many students are concerned about how their math backgrounds have prepared them for statistics. Math background is important because if one's math background is weak, then valuable time is spent learning how to do the math in the formulas instead of understanding what the statistics produced by the formulas mean.

I have no doubt that you can add, subtract, multiply, and divide, but you may be out of practice if you have not computed recently. Similarly, it may have been awhile since working with positive and negative numbers, rounding to the hundredth, squaring a number and taking its square root, working with fractions, and solving algebra problems. In your statistics course, you will need to do all of these. This introductory chapter is designed for review and practice with the computations you will encounter throughout the course. A diagnostic test concludes the chapter with an answer key located in the back of the book.

How well do you rate yourself doing math? Weaver, Drone, Varela, and Heskett (1995) in their research based on 150 statistics students reported that students' response to this question was the best predictor of how well students did in the statistics course. If you rate yourself low in doing math but have completed the college algebra course, you have the background to successfully do the computations in your statistics course. If you rate yourself low AND have not yet had college algebra, then wait to take the statistics course until you have completed college algebra.

Regardless, Weaver et al. reported that math anxiety greatly diminished from the beginning to the end of the course. Isn't this a strange result when you look at the formulas just before this chapter and see how they become more complex? Not really. Once you have practiced and reviewed the computations in this chapter, you will be ready for the math in this course.

Assistance. Knowing where help is available when you are experiencing difficulty with any material in your statistics course reduces math anxiety. Usually, you will have several options for seeking assistance. You can meet with your professor, ask questions in class, arrange a session with a tutor if one is provided, visit your campus math lab if your campus has one, seek help from a fellow student in the course who understands the material and is willing to assist you, become part of a study team with fellow students in the course, or identify a student (someone in your residence hall or fraternity or sorority) who has taken the course and done well and is willing to assist you. The statistics course unfolds quickly and is full of fascinating and useful content that sequentially builds on itself, becoming more complex as it does so. Seeking assistance immediately when you do not understand a topic as future course content will require your understanding of current content.

Calculator. If you need a calculator for the course, a calculator you can use easily and accurately will help ease anxiety. A fancy or graphing calculator may not be necessary (check with your instructor), rather find one that is reliable and easy to use. Put new batteries in your calculator to ensure its performance throughout the course and especially for tests. You will use the calculator for adding, subtracting, multiplying, dividing, squaring, and finding the square root. A key on the calculator that makes values positive or negative is useful. Calculators which have statistics functions or which graph are not necessary for the introductory course, unless your instructor informs you differently. A good calculator can be purchased for $10 to $15 and is a good investment.

Statistical software. Some textbooks include a computer disk containing statistical software to use for solving problems instead of a calculator. Your instructor may require you to use the disk or to use a large statistical software package like SPSS or SAS. You want to become familiar with these tools early in the course so you can devote most of your time to understanding the statistics.

Course Objectives

What goals should you have for the course? The following are good ones to accomplish:

1. To gain factual knowledge about statistics.
2. To learn fundamental principles of statistics.
3. To begin becoming acculturated to the statistical perspective.
4. To understand the reasoning behind the statistical procedures covered in the course.
5. To realize that there is always a probability of being wrong when making a conclusion based on data. All that researchers can do is attempt to make this probability as small as possible.
6. To understand the relationship among the null hypothesis, statistical testing, and the scientific method.
7. To appreciate that statistics is not an end in itself, rather it is a tool that when used properly assists researchers' quest for truth. Like any tool, abuse or improper use may be harmful.
8. To become more comfortable and proficient with computation.
9. To connect statistical concepts of normality and variability with social constructs of normality and variability.
10. To interpret the statistical section of a research article.

Student Learning Outcomes

If your instructor goes as far as analysis of variance and correlation, the following outcomes are among the competencies you should strive to learn by the end of the course and form the core of this book:

1. Differentiate among nominal, ordinal, interval, and ratio scales of measurement.
2. Compute and explain the mean, median, mode, range, sum of squares, variance, standard deviation, z score, standard error of the mean, standard error of the difference, standard error of estimate, t score, F, r, and coefficient of determination.
3. Correctly read and write the symbolic notation of statistics.
4. Link types of kurtosis with descriptions of homogeneity and heterogeneity and with corresponding fluctuations in the magnitude of the standard deviation.
5. Relate null and alternative hypotheses to the research hypothesis or the research question and the particular statistical test.
6. Discriminate among different types of sampling distributions by using them properly.
7. Apply inferential statistical procedures correctly in order to answer research questions.
8. Interpret computer printouts of descriptive statistics, the t test, analysis of variance, correlation, and simple regression.

Exercise

Using a highlighter, mark the sections of the passage below which you do not understand. This same exercise is repeated at the end of Chapter 10 to demonstrate to you how much you will have learned in a comparatively short time. Begin highlighting until you reach the instruction to stop:

Rejecting the null hypothesis requires that the calculated value of F exceeds the critical value of F coming from the table. Rejecting the null hypothesis indicates that the means of the three treatment conditions are different from one another. If the scientific hypothesis predicts that the treatment conditions will be different from one another, then the size of the calculated value of F should be maximized in order to reject the null hypothesis. From looking at the formula for the F ratio below, how can the size of the F be increased?

$$F = \frac{\dfrac{(M_1 - M_t)^2 n_1 + (M_2 - M_t)^2 n_2 + (M_3 - M_t)^2 n_3}{k-1}}{\dfrac{\Sigma(X_1 - M_1)^2 + \Sigma(X_2 - M_2)^2 + \Sigma(X_3 - M_3)}{N-k}}$$

Regard this formula for the F as being three fractions: the top fraction, the bottom fraction, and the overall fraction. Thus, there is the numerator of the numerator, the denominator of the numerator, the numerator of the denominator, and the denominator of the denominator.

Maximizing F occurs when there is heterogeneity between groups and homogeneity within groups. Thus, the F is increased if the size of the numerator of the numerator is increased (i.e., greater heterogeneity between groups) and the size of the numerator of the denominator is decreased (i.e., greater homogeneity within groups). In addition, F is increased as the sample size increases, which is reflected in the denominator of the denominator.

Stop! How much of this passage did you highlight? I would guess most or all of it. By the time you finish Chapter 10 The One–Way Analysis of Variance, you will know this material and much, much more.

Review

Integers and Even and Odd Numbers

An integer is a whole number like 8, 32, and 629. In contrast, $\frac{2}{3}$, 43.63, and .91 are not whole numbers. An even number is any number that is divided by 2 AND produces an integer for an answer. 2, 4, 6, 8, and 10 are all even numbers. An odd number is one that will not produce an integer for an answer when it is divided by 2. 1, 3, 5, 7, and 9 are examples of odd numbers.

The Number Line and Positive and Negative Numbers

On a number line, positive numbers are to the right of 0, and negative numbers are to the left of 0. You will see throughout the course these associations between being positive and the right side and between being negative and the left side. On the number line, a number located to the right of any other number will always be larger. A number located to the left of any other number will always be smaller. For example, 3 is larger than 1 and –1 is larger than –3. Conversely, –4 is smaller than –2, but 2 is smaller than 4.

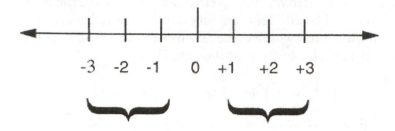

To the left of 0   To the right of 0
is the negative   is the positive
side   side

Looking Ahead

You will associate left with negative and right with positive throughout the course. In Chapter 3, you will learn about positive and negative skewness. In Chapter 5, you will compute + and - z scores. In Chapter 6 you will learn about the left and right tails of the normal distribution. In Chapters 7, 8, and 9, you will use the right and left sides of sampling distributions.

Greater Than and Less Than

Math uses many symbols. The +, –, x, ÷, and = signs are examples. Some symbols represent *inequality* such as not equal to, greater than, less than, greater than or equal to, and less than or equal to. What does inequality mean? It means not equal to or not the same as. It means that whatever is on one side of the inequality symbol is *different* from what is on the other side of the inequality symbol.   For example, $8 \neq 10$.

It is possible to be more specific when talking about inequality. Two students stand up, one is 5'8" and the other is 6'1". You can say that the two students' heights are different. You can also specify the *direction* of that difference and say that the 6'1" student is taller than the 5'8" student. While $\neq$ means *different from*, > is the symbol for *greater than* and < is the symbol for *less than*. Thus,

5 $\neq$ 4, but 5 > 4 is more specific and means "5 is greater than 4"

–2 $\neq$ –3, but –2 > –3 is more specific and means "–2 is greater than –3"

$4 \neq 5$, but $4 < 5$ is more specific and means "4 is less than 5"

$-2 \neq -3$, but $-3 < -2$ is more specific and means "–3 is less than –2"

Think of the $<$ and $>$ symbols as arrow tips pointing to the *smaller* of the two values. So

$5 \rightarrow 4$, thus $5 > 4$. Or $4 \leftarrow 5$, thus $4 < 5$.

In addition, there are the symbols $\leq$ and $\geq$, which mean *less than or equal to* and *greater than or equal to*, respectively.

Looking Ahead

In Chapter 11, the range of values for the correlation will be given as $-1 \leq r \leq 1$. This expression means that the correlation is a value which falls between -1 and 1, including being -1 or 1.

Fractions

If you have any "ills" about working with fractions, the statistics course will surely cure you. Almost all of the formulas are written in fraction form. The fraction is made up of three parts: the numerator which is the number on top, the denominator which is the number on the bottom, and the horizontal line which means "divided by." So, $\dfrac{numerator}{denominator}$ = numerator divided by denominator = denominator divided into the numerator = $denominator\overline{)numerator}$

Thus, $\dfrac{18}{6} = 6\overline{)18} = 3$, and $\dfrac{6}{18} = 18\overline{)6} = .33$. To divide using your calculator, press the keys for the number in the numerator, the $\div$ key, the keys for the denominator, and finally the $=$ key. Your answer appears in the calculator's window.

Mnemonic: A good technique for remembering that the numerator goes on top and the denominator is placed on the bottom is to match the *n* of numerator with the "n" of north (and north is usually at the top of maps) and the *d* of denominator with the "d" of down.

- Whenever the numerator is greater than the denominator, the overall value of the fraction is greater than 1.
- Whenever the numerator is equal to the denominator, the overall value of the fraction is 1.
- Whenever the numerator is less than the denominator, the overall value of the fraction is less than 1.

The final answers to problems no longer appear in fraction form. If you end up with a fraction, perform the division to produce the actual number. Thus, $6\frac{1}{2}$ is reported as 6.50.

Adding and Subtracting Fractions. Computing with fractions can be tricky. One approach is to use your calculator to convert the fraction to a number. Sometimes, though, working with fractions is more efficient. *Adding and subtracting fractions can only be done when the denominators of the two fractions are the same.* If the denominators are not the same, you have to find the least common denominator. The least common denominator is the smallest number that both of the denominators can be divided into and produce an integer as the quotient. Regardless of the approach you use, no more fractions appear in the final answer.

When the denominators are the same, then you can add or subtract the numerators. Thus, $\frac{1}{2}+\frac{1}{3}=\frac{3}{6}+\frac{2}{6}=\frac{5}{6}$. You then divide 6 into 5 (that is, $5 \div 6$ or $5\overline{)6}$) with your calculator. The quotient rounded to the hundredth is .83 and rounded to the thousandth is .833.

Look at the following subtraction problem: $\frac{1}{2}-\frac{1}{3}=\frac{3}{6}-\frac{2}{6}=\frac{1}{6}$. Dividing the 6 into 1 produces a quotient of .17 rounded to the hundredth and .167 rounded to the thousandth.

Multiplying Fractions. When multiplying two fractions together, you multiply numerator by numerator and denominator by denominator. Thus, $\frac{1}{2} \times \frac{1}{3}=\frac{1 \times 1}{2 \times 3}=\frac{1}{6}$, which rounded to the hundredth is .17 and rounded to the thousandth is .167.

Dividing Fractions. When dividing two fractions, the rule is to invert the denominator and multiply. *Critical Thinking Question*: What does invert mean? Turn upside down.

Thus, look at what happens to $\frac{1}{3}$ in the denominator below:

$$\frac{1}{2} \div \frac{1}{3}=\frac{\frac{1}{2}}{\frac{1}{3}}=\frac{1}{2} \times \frac{3}{1}=\frac{3}{2}=1.50$$

Alternately, you can first convert the fractions into decimals and then divide. Thus, $\frac{1}{2} \div \frac{1}{3}$ becomes $.50 \div .33 = 1.52$. The difference between 1.50 and 1.52 is due to rounding $\frac{1}{3}$ to .33.

Adding and Subtracting Positive and Negative Numbers

The answer in addition is called the what? The sum. The answer in subtraction is called the what? The difference.   When working with positive numbers, addition and subtraction are straight forward. However, when working with negative numbers, these operations become more complicated. For example, $3 - 2 = 1$ is the same as $3 + (-2) = 1$. Subtracting a positive number is computationally the same as adding a negative number. Here are some illustrations:

1. Adding a positive number is the same as addition:
    $+12 + (+6) = 12 + 6 = 18$
    $-12 + (+6) = -12 + 6 = -6$

2. Subtracting a positive number is the same as subtraction:
    $+12 - (+6) = 12 - 6 = 6$
    $-12 - (+6) = -12 - 6 = -18$

3. Adding a negative number is the same as subtraction:
    $+12 + (-6) = 12 - 6 = 6$
    $-12 + (-6) = -12 - 6 = -18$

4. Subtracting a negative number is the same as addition:
    $+12 - (-6) = 12 + 6 = 18$
    $-12 - (-6) = -12 + 6 = -6$

Stories illustrate these combinations. For example, you have no money, and I loan you $6. How much money do you now have? You have $-$$6. You then do a chore for which you are paid $6. If you pay me back, the computation is the $6 you earned plus the $-$$6 loan you repay, which is $6 + (-$6)$, which is $6 - 6, which equals $0 left.

If I tell you that you do not have to repay the loan, then how much money do you now have? You have the $6 you earned minus the $-$$6 loan (canceling the $6 loan is translated numerically as "subtracting the $-$$6 loan). The computation is $6 - (-$6)$, which equals $6 + 6, which equals $12.

Remember that you can add 0 to any number or subtract 0 from any number without changing that number.

```
┌─────────────────────────────────────────────────┐
│                  Looking Ahead                   │
│ In Chapter 9, you will be subtracting 0 from a   │
│ number in the t formula for the two-sample t test . │
└─────────────────────────────────────────────────┘
```

Multiplying and Dividing Positive and Negative Numbers

The answer in multiplication is called the what? The product. The answer in division is called the what? The quotient. Multiplying and dividing with positive and negative numbers has its own set of rules. They can be demonstrated as follows:

1. Multiplying two positive numbers together will produce a positive product:
   $+12 \times +6 = 12 \times 6 = (12)(6) = 72$

2. Multiplying a positive and a negative number together will produce a negative product:
   $+12 \times -6 = 12 \times -6 = (12)(-6) = -72$

3. Multiplying two negative numbers together will produce a positive product:
   $-12 \times -6 = -12 \times -6 = (-12)(-6) = 72$

4. Dividing two positive numbers will produce a positive quotient:
   $+12 \div +6 = 12 \div 6 = 2$

5. Dividing a positive and a negative number will produce a negative quotient:
   $+12 \div -6 = 12 \div -6 = -2$

6. Dividing two negative numbers will produce a positive quotient:
   $-12 \div -6 = 2$

Multiplying or dividing a number by 1 produces that number as the product or quotient.

Rounding to the Hundredth

This book follows the convention of rounding numbers to the hundredth. Your calculator's answer will have more than two digits to the right of the decimal. If so, you will need to round. Consider the number 2126.38517:

The thousandth column
↓

$$2126.38\overset{}{5}17$$

↑
The hundreth column

To round to the hundredth, look at the digit occupying the thousandth column, which is the third column to the right of the decimal. If the digit in the thousandth column is between 0 and 4, then the number in the hundredth column remains as is. However, if the digit in the thousandth column is between 5 and 9, then the number in the hundredth column is increased by one. For example, 2126.38517 has 5 in the thousandth column. To round this number to the hundredth, the 8 in the hundredth column increases by one, and the number is written as 2126.39.

Solving for an Unknown

This course is full of formulas in fraction form, which will require you to determine the value of something unknown. You are probably most familiar with the term "solve for X" where X symbolizes what is not known. In this course, you will encounter problems where the mean is unknown, or the standard deviation is unknown, or the z score or t score is unknown.

When solving for an unknown, the challenge is to isolate the unknown by itself on one side of the = sign. This means that all the computations for solving the unknown are located on the other side of the = sign.

$5 = \dfrac{15}{3}$ correctly expresses a true relationship among the three numbers 3, 5, and 15. What if the formula were $X = \dfrac{15}{3}$, and your task was to solve for X. How would you do this? You could say "Oh, that is so easy. I can just look at the problem and know the answer is 5." But the problems you will see throughout the course are not going to be this easy.

You need to know what steps you followed to come up with the answer of 5 so you can apply those same steps when you are solving more difficult problems. In this example, the unknown is already by itself on the left side of the =; all of the computations are on the right side. Thus, $\dfrac{15}{3}$ is the same as $3\overline{)15}$, which is the same as $15 \div 3$, which equals 5.

Now what about $5 = \dfrac{X}{3}$? What are the steps necessary to isolate X by itself on one side of the equation? First, let's review the main rule: *You can do anything to one side of the equation as long as you do the exact same thing to the other side.* By following this rule, the meaningfulness and integrity of the = is always maintained. In this example, you first multiply both sides by 3. Thus, you have $(3)(5) = \dfrac{X}{3} \times \dfrac{3}{1}$ which becomes $15 = X$.

Finally, there is $5 = \dfrac{15}{X}$. What steps do you follow to isolate X by itself on one side of the equation? What do you do first? The first step is to multiply both sides of the equation by X. Thus, you have $(X)(5) = \dfrac{15}{X} \times \dfrac{X}{1}$, which becomes $5X = 15$. Then you divide both sides by 5 like so: $\dfrac{5X}{5} = \dfrac{15}{5}$. This equation simplifies to $X = 3$, and 3 is your answer.

Arithmetic and Algebra Diagnostic Test

For Problems 1 through 22, solve for X.

1. $14.18 + (-3.69) = X$

2. $600{,}543.21 - 129{,}984.65 = X$

3. $65{,}289.35 - 63{,}412.84 = X$

4. $53.80 - (-16) = X$

5. $42.37 - X = 55$

6. $\dfrac{150 - X}{10} = 8.16$

7. $\dfrac{15}{326} \times -\dfrac{241}{14} = X$

8. $258 \times X = -63$

9. $\dfrac{43 + 21 + 65 + 23 + 49 + 82}{6} = X$

10. $-17 \times 1.68 = X$

11. $83.25 - 124.33 = X$

12. $\dfrac{348}{16} = X$

13. $255.61 = \dfrac{3154.99}{X}$

14. $56.21 = \dfrac{X}{1.18}$

15. $\dfrac{8}{15} \div \dfrac{23}{13} = X$

16. $\dfrac{3}{8} + \dfrac{4}{7} = X$

17. $X = 18.23^2$

18. $X = 1294^2$

19. $X = .47^2$

20. $X = -6.98^2$

21. $X = \sqrt{289.66}$

22. $X = \sqrt{.74}$

For Problems 23 through 30, indicate true or false. N stands for numerator, D stands for denominator.

23. If N < D, then $\dfrac{N}{D} < 1$

24. If N > D, then $\dfrac{N}{D} > 1$

25. If N = D, then N − D = 1

26. If N = D, then D − N = 0

27. If N = D, then N + D = 2 x N

28. If N = D, then N + N = 2 x D

29. If N = D, then $\dfrac{N}{D} = 1$

30. If N > D, then $\dfrac{N}{D} < 1$

31. Circle all of the integers: 8   16.25 $\frac{3}{4}$ 121.00   3.85   .50

For Problems 32 through 36, round the numbers to the hundredth.

32. .2459999

33. 6.8317

34. 754.935

35. 18.9393

36. 25643.344999999

For Problems 37 through 42, indicate true or false.

37. 8.13 > 4.28

38. −4 < −6

39. 7.35 > −8.42

40. −1 < 1

41. A number divided by itself equals 0.

42. Two negative numbers multiplied together will produce a negative quotient.

Chapter 2

Measurement and Vocabulary

Thinking of Numbers: Scales of Measurement

If your car does not have gas, then it will not run. If your formula does not have numbers, then it will not run. Numbers are essential in statistics. Stevens (1951) introduced a useful way of thinking about what numbers measure and what can be done with them. He proposed that a number belongs to one of four scales of measurement: nominal, ordinal, interval, and ratio. A "rule for the assignment of numerals (numbers) to aspects of objects or events creates a *scale*" (p. 23). What rules differentiate these four scales from one another?

Nominal Scale. In the nominal scale, numbers are used as names or as categories.   Your student identification number belongs to the nominal scale of measurement. The #22 on a softball player's jersey belongs to the nominal scale of measurement. If Freshman on a survey is coded as 1, Sophomore as 2, Junior as 3 and Senior as 4, the four numbers are used to represent categories and belong to the nominal scale of measurement.

You cannot say a Senior is four times better than a Freshman because the Senior was coded as 4 and the Freshman was coded as 1. You cannot say that the athlete wearing #44 is twice as good as the athlete wearing #22. Adding, subtracting, multiplying, or dividing these nominal numbers does not make sense.

You are limited in what you can do with numbers from the nominal scale. For example, you can count how many seniors responded to the survey by counting the number of 4s. Using the chi-square test in Chapter 12 enables you to evaluate the actual patterns of frequencies found in the data with the frequencies you expected to obtain. That is about the only analysis you can perform with nominal data.

Ordinal Scale. Numbers on the ordinal scale represent ranks. You know that one rank is different from another rank, but you cannot determine by how much. For example, the first place float in the homecoming parade was built by a fraternity, the second place float was built by a residence hall, and the third place float was built by a student organization. The ranks tell you that the fraternity float was the best but you do not know how much better the fraternity float was in comparison to the residence hall float. Likewise, you do not know how much better the residence hall float was in comparison to the student organization's float. The three could have been very similarly rated. Or perhaps the first place float was much better than the other two. Knowing only the ranks limits how much you know about the difference between the ranks.

Interval Scale. Unlike the ordinal scale, numbers on the interval scale represent the same amount of difference. For example, the difference in temperature between 44° and 45° is the same as the difference between 84° and 85°. For the homecoming parade, you learn that out of 100 possible points (50 for creativity and 50 for the theme), the judges awarded the fraternity float 95 points, the residence hall float 90 points, and the student organization float 70 points. By knowing the number of points each float received, you can see that the competition for the best float was keen while the third place float was not nearly as good as the first two. Knowing the floats' scores provided richer meaning than knowing just their ranks.

Ratio Scale. What is a one word synonym for ratio that begins with the letter *f*? A fraction is a ratio. Knowing that a fraction is a ratio is helpful in understanding how a ratio scale of measurement differs from an interval scale. For example, Jan's resting heart rate is 50 beats per minute. During a cardiac stress test , Jan's heart rate elevates to 100 beats per minute. Can you say that Jan's heart is beating twice as fast? Yes: $\frac{100}{50} = \frac{2}{1} = 2$ or twice as fast. What underlies using fractions is a scale of measurement which has a *meaningful* 0 that actually means 0. For beats per minute, there is a meaningful 0 when someone has died.

Temperature is a measure of heat. If the outside temperature was 40° F yesterday and 80° F today, is it twice as hot today as yesterday? No. And why isn't 80° F twice as hot as 40° F? Because 0° on the Fahrenheit temperature scale, just like 0° on the Celsius temperature scale, is not a meaningful 0. 0° Fahrenheit or 0° Celsius does not mean the absence of all heat.

There is a temperature scale, however, which does have a meaningful 0. Do you know what the name of that temperature scale is? Kelvin. Because 0° Kelvin is absolute zero, you can say that 100° Kelvin is twice as hot as 50° Kelvin.

Here is another example. Sally scores 90 and Jill scores 45 on their first statistics test. Based on the scores, can you conclude that Sally is twice as smart as Jill? No. Why not? Because you cannot say that a 0 on the statistics test means the absence of all knowledge about statistics. Can an instructor create a test on which you make a 100? Sure. Can an instructor create a test on which you make a 0? Sure. What can you conclude from this is that the way a test is constructed can contribute to the score. A test score of 0 means that the test may not have been sensitive enough to detect the student's knowledge in the area. Test scores and most other measurements in psychology and education belong to the interval scale of measurement. In contrast, most measurements in nursing and medicine belong to the ratio scale of measurement where death is the meaningful 0 (no respiration, no blood pressure, no heart rate, no ankle-brachial index, etc.).

What are other examples of ratio scale measurements? Height, weight, distance, and time are some examples. The ratio measurements in a medical setting include heart rate, visual acuity, blood pressure, and respiration rate. *Critical Thinking Question*: What does the meaningful 0 mean for respiration rate? The person is not living.

Terminology

During the course, your instructor and your textbook will use a variety of statistical terms. Here is a primer on the important ones.

Before any measurements are collected, the researchers have to identify what their population is and what their sample is. Defining the population can be tricky. Population does not just mean all the people in the city or all the people in the state or all the people in the nation. Rather, the *population* is defined to be whatever the researchers need it to mean. A researcher could define the population as all the companies earning more than $1 million in states east of the Mississippi River. Or all the psychology majors at the state universities in Kansas. Or all the ants living between the 40 and 50 yard line on the football field. Or the students enrolled in this semester's statistics course. Or the hospitals with less than 100 beds in Colorado. The researcher defines the population, then the sample is defined as a subset of the population. The descriptors of the population, such as its mean and standard deviation, are called *parameters*.

Defining the *sample* is the easy task; it is any subset of the population. A random sample is any subset of the population whose members all had an equal likelihood of being selected to be part of the sample. The descriptors of the sample such as its mean and standard deviation are called *statistics*.

Once the population and sample are identified, the researcher is ready to start collecting information from the research participants. This information can be both qualitative (e.g., opinions on a topic, relating an eyewitness account) and quantitative (e.g., when the participants in the research have been measured in some way). This information is called *data* (the plural of datum). For this course, you use quantitative data or measurements gathered usually from a sample, but sometimes you use population data (e.g., the US Census).

The measurements are often scores from a test or task the participants have completed. The number of points on a test, the number of words remembered from a list of 20, the amount of time a rat runs from the start box to the goal box, the performance on the cardiac stress test, and the distance one throws a softball are examples. *Scores* may also be personal characteristics such as one's weight or height or blood pressure. Scores may also be referred to as *raw scores*. The reason for needing the adjective "raw" to qualify score is that later in the course, you encounter deviation scores and *z* scores and *t* scores. If someone just says scores, the context may not be specific enough to know if the speaker is talking about raw scores, deviation scores, *z* scores, or *t* scores. Thus, the adjectives become crucial.

After the data are collected, the researchers end up with a bunch of numbers. Their first task is to order the data by arranging them in numerical sequence. With this arranging, the bunch of numbers is converted into a *distribution* or *data set*.

The presentation of statistics in this book follows the traditional sequence of presenting descriptive statistics first and then inferential statistics. The goal of *descriptive statistics* is to understand the group of numbers that is your sample or population. Descriptive statistical procedures reveal characteristics of the distribution such as the location of its center (the mean, median, and mode) and the spread of scores from that center (the sum of squares, variance, and standard deviation). The goal of *inferential statistics* is to determine how accurately the descriptive statistics of the sample describe the population from which the sample was selected.

Looking Ahead

What is it called when a friend sets you up with a date? A blind date. This phrase does not mean that the date is blind, but rather you do not know anything about this person; you are "blind" to the person until you meet the person and become familiar with her or him. Similarly, until the researchers start using statistical procedures on a bunch of numbers, they do not know anything about their data. They are "blind" to their data. Once they start using the procedures in Chapters 3, 4 and 5, they come to know their data through these procedures just as you come to know your blind date through the questions you ask and the conversation you share.

Problems

1. What differentiates a data set from a bunch of numbers?

2. How do descriptive and inferential statistics differ?

Determine whether the statements in Problems 3 through 8 are true or false.

3. If the population is made up of all the even numbers between 0 and 100, then 18, 52, and 65 constitute a sample.

4. It is twice as warm when it is 60° Fahrenheit as when it is 30° Fahrenheit.

5. If you score a 90 on your test and I score a 45, you are twice as smart as I am.

6. If you score a 90 on your test and I score a 45, you know the material twice as well as I do.

7. If you score a 90 on your test and I score a 45, you have twice as many points on your test as I do.

8. Your social security number belongs to the nominal scale of measurement.

9. Runner #88 finished the 5 mile race in first place with a time of 26.50 minutes. Indicate the scale of measurement for all numbers in this sentence.

10. During the game, Suzy shot the basketball 10 times, and Shonda shot the basketball 15 times. What scale of measurement does the number of basketball shots belong to?

Chapter 3

The Distribution: Finding Its Center and Eyeballing Its Shape

To answer a research question or test a research hypothesis, you identify a group of organisms to study. The organisms could be college students at a university, all the third graders in a school district, companies in eight midwestern states which employ less than 300, hospitals with 200 or more beds, or the ducks which occupy a particular pond.

To collect the data you need from the members of your sample or population, you measure them in some way and end up with a bunch of numbers. Now what do you do? You apply different statistical procedures to your data to understand them, just as you ask questions to know your blind date. These procedures can be loosely referred to as "massaging your data."

The first statistical procedure is to arrange your scores in numerical order. This contact with your data transforms that "bunch of numbers" into a data set or distribution.

The next procedures commonly applied to a distribution are to find its center. These procedures are attempts to identify the most characteristic number in the data set. There are three different descriptors for the center of a distribution: the mean, the median, and the mode.

The Mean

The sample mean is frequently the first formula you encounter in the statistics course. As you may expect from the Chapter 1 Arithmetic and Algebra review, it is in the form of a fraction:

$$M = \frac{\Sigma X}{N}$$

How many times have you computed the mean? Lots. Please bear with me a moment as you formally tour this rather simple equation. Every part of it is important.

The formula expresses the relationship among three variables—M, ΣX, and N using symbols to do so. Note that one of these variables, ΣX, contains two symbols. Your recognizing this is crucial for successful computing and understanding later in the book. Let's dissect the formula, being just as systematic as you would be with a frog in a biology class.

The M is the symbol for the sample mean, Σ is the symbol which means to add, X is the symbol for a raw score, and N is the symbol for the number of scores in the data set. The combination ΣX represents one number and is the *sum* of all of the scores in the data set. Although there are four symbols, the three variables are the sample mean, the sum of the raw scores, and the number of raw scores in the data set.

To compute the mean requires you to divide the sum of the raw scores by the number of raw scores in the data set and round your answer if necessary. Piece of cake computationally, but what do you really know when you know the mean?

The mean gives you two very important pieces of information: the mean is the average raw score (or the average) AND the mean is the *value which produces the smallest sum of all of the squared deviation scores* (more about deviation scores in the next chapter). No value other than the mean can lay claim to these two qualities. That's why the numerator of the formula for the mean is so critical, because computing ΣX requires the value of every score in the sample be included in the computation to find the mean.

The formula for the mean can be rearranged to isolate the other two variables on one side of the equation. This is useful in case you want to compute the ΣX or N, and later in this chapter an example is given where you have to compute ΣX from M and N. The two variations of the formula for the mean are:

$$\Sigma X = M \times N \quad \text{and} \quad N = \frac{\Sigma X}{M}$$

Critical Thinking Question: Do you know what has been done algebraically to transform the formula for the mean into these two variations? Hint: You can perform any computation to one side of the equation as long as you perform the same computation to the other side.

The Median

The median of the interstate highway is the strip of land or barrier dividing the traffic moving in one direction from the traffic moving in the opposite direction. The median divides the highway in half such that one–half of the highway is on one side of the median, and the other half of the highway is on the other side of the median. Similarly, the median of a data set is the value which divides the data set in half. Thus, one–half of the scores are greater than the median and one–half of the scores are less than the median.

The median is computed one way if the number of scores in the data set is odd and a slightly different way if the N is even. Note that finding the median requires the scores to first be in numerical sequence. If a data set has an odd N, the median is the score located in the $\frac{N+1}{2}$ position. A sample of five adults' ages is 38, 51, 55, 68, and 79. What is the median of this data set? The N is 5, so 5 + 1 equals 6, and 6 divided by 2 equals 3.

CAUTION: *The median is not 3.* Think of what 3 means for this data set—the median age of five adults is 3 years of age??? This does not make sense. Rather, 3 is the *position* in the data set occupied by the median. Whether you start counting from 38 and go forward or start with 79 and count backward, the median is 55. Two scores in the data set are above the median, and two scores in the data set are below the median.

38   51   55   68   79

↑

Median

Suppose that a sixth adult's age of 83 becomes part of the data set. Now the N is an even number. How do you find the median? When the N is even, the median is the value halfway between the scores occupying the $\dfrac{N}{2}$ and $\dfrac{N}{2}+1$ positions. Thus, the median is halfway between

the scores occupying the 3rd and 4th positions, or $\dfrac{55+68}{2}=\dfrac{123}{2}=61.50.$

$$61.50$$

38   51   55   68   79   83

⬆

Median

The Mode

The mode is the score in the data set which occurs most frequently. The mode of the distribution is recognizable because it is the tallest point in any frequency distribution graph.

Skewness

You are familiar with what is called the normal "bell shaped" curve. The curve is the result of a mathematical formula and represents an ideal distribution. However, many distributions approximate the normal shape and are therefore considered to be normal.

A graph has two axes, the Y axis, which runs vertically, and the X axis which runs horizontally. The name of the Y axis is the *ordinate*, and the name of the X axis is the *abscissa*. For the normal curve, the scores are located on the abscissa, and the number of times each of the scores occurs in the distribution is located on the ordinate. Often, the normal distribution is presented without the Y axis, but it is understood to be there. Here is the normal distribution:

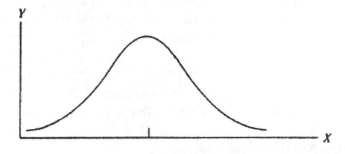

The normal curve is considered to be symmetrical or balanced, which means that one–half of the scores in the sample are below the mean, and the other one–half of its scores are above the mean. Also, the curve is unimodal—it has one mode which you see as one "hump" in the distribution. Can you imagine what a bimodal distribution looks like?

The normal curve has 0 skewness because it is symmetrical—one–half of the scores are on one side of the mean, and one–half of the scores are on the other side of the mean. The two distributions below are not symmetrical. The curve on the left is skewed *negatively*, and the curve on the right is skewed *positively*. What does it mean for a distribution to be skewed, and how does a distribution come to be skewed either positively or negatively?

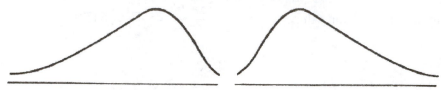

A normal distribution becomes skewed, or distorted, when most of the scores are in the central section of the distribution but a few extreme scores reside in one of the tails of the distribution. In a negatively skewed distribution, more unusually low scores are located in the left tail of the distribution than expected by the normal distribution. In the positively skewed distribution, more unusually high scores are located in the right tail than expected by the normal distribution.

In the normal curve, the mean, median, and mode are the same value and located in the center. In a skewed distribution, these three measures of central tendency are different values and thus are not located in the same position.

For both normal and skewed distribution, the mode is always the tallest point of the distribution. *Critical Thinking Question:* Do you remember why the mode is the tallest point of the distribution? Because the mode is the score which occurs most frequently. The mode then is the easiest descriptor to locate; it is the score on the abscissa corresponding to the tallest point.

Where is the mean of a skewed distribution located on the abscissa? Calculating the mean requires all of the scores in the distribution. The unusually low scores in the negatively skewed distribution keep the mean small. Like a magnet, the low scores relative to the rest of the scores keep the mean smaller than the median, moving the mean to the left of center and closer to the unusually low scores. Conversely, in the positively skewed distribution, the unusually high scores keep the mean larger than the median, moving the mean to the right of center and closer to them. With either type of skewness, the mean loses its ability to define the distribution's center.

Where is the median in a skewed distribution? The median regards the scores as occupying various positions. Determining the median requires using the *N* of scores but not their value. Unlike the mean, computing the median does not require adding the scores together. Consequently, the median is not as susceptible as the mean is to the extreme scores. In a skewed distribution, the median is the best measure of central tendency.

Example

A magazine advertisement for the Army several years ago reported that two–thirds of new recruits scored above the mean on the Armed Forces Qualification Test, a standardized aptitude test. Think about the distribution of the new recruits' scores. You know it will be skewed because two–thirds of the recruits scored above the mean on the test. In a normal distribution, 50% of the scores are above the mean. Where are the unusual scores, in the left tail or the right tail? Left tail.

Is the distribution skewed negatively or skewed positively?   Negatively. Is the mean located toward the left or right side of the distribution? Left

The Frequency Distribution Table and Frequency Polygon

Once you know the mean, median, and mode, you are ready to convert the scores in your data set into a frequency distribution polygon. This visual representation of your data set can reveal some very interesting information such as whether you have any outliers (very unusual scores) or how much your distribution looks like a normal distribution.

Below is a data set of adult heights in inches from one of my statistics classes a few years ago. This class had 41 students, 28 women and 13 men. The numbers are already in numerical order so it is appropriate to call this sample a distribution.

Women's Heights					Men's Heights	
78	68	66	62		72	69
71	68	66	61		72	68
70	67	66	60		72	68
69	67	65	59		72	68
68	67	64	56		71	68
68	67	64	53		71	65
68	66	63	50		71	

The first step is to compute the mean, median, and mode for the set of 41 measurements. Please do this and round your answers to the hundredth: $M = 66.44$, median = 68, mode = 68

Creating the Frequency Distribution Table

Next compute the *inclusive range*, which is the highest score – the lowest score + 1. The *exclusive range* is the highest score – the lowest score. For the 41 scores, 78 is the highest score and 50 is the lowest score so the inclusive range is $78 – 50 + 1 = 29$ inch range.

The second step is to decide how many intervals you want your frequency distribution table to have. The number must be between 10 and 15. You will then divide your number of intervals into your range to find the size of each interval. The interval size must be an integer so selecting a number of intervals that divides into the range to produce an integer quotient is good. If not, then *you will always round the quotient up.* So, 10 intervals $\overline{)29}$ inch range = 2.90 inches/interval, which we round up to become an interval size of 3.

Once you have your interval size, you are ready to start constructing your intervals. The preference is for starting with the smallest score and working up, but you can also start with the largest value and work down. The smallest height is 50" so 50 is where the first interval begins.

Critical Thinking Question: If the interval size is 3 and the first interval begins with 50, what score will end the first interval? 52. You may be thinking that $50 + 3 = 53$, so the first interval is

50–53. But this interval actually contains four heights—50, 51, 52, and 53—and has an interval size of 4 instead of 3. So the correct first interval is 50–52. Now create the rest of your intervals.

The intervals are located in the first column in your table. The other two columns contain Frequency and Relative Frequency. For each interval, indicate in the Frequency column how many scores in the data set belong in that interval. Then divide each frequency by the N and put the answer (rounded to the hundredth) in the Relative Frequency column.

Once your frequency distribution table is complete, you can check for errors by adding the frequencies and adding the relative frequencies. The frequencies should sum to be the N. If not, then you have missed a score or counted a score twice. The relative frequencies should sum to be 1. If the sum is slightly off (i.e., between .97 and 1.03), then you are probably seeing the cumulative effect of rounding most of the relative frequencies either up or down. This effect is called *rounding error*. The frequency distribution table for the 41 scores looks as follows:

Interval	Frequency	Relative Frequency
77–79	1	.02
74–76	0	.00
71–73	8	.20
68–70	12	.29
65–67	10	.24
62–64	4	.10
59–61	3	.07
56–58	1	.02
53–55	1	.02
50–52	1	.02
	$\Sigma X = 41$	$\Sigma rf = .98$

Graphing the Frequency Polygon

Once the frequency distribution table is in place, then you are ready to construct your graph of the distribution. There are several ways to visually present your data such as a histogram, pie chart, or stem and leaf display. Here is what a computer–generated stem and leaf display of a data set of 44 women's weights looks like:

Frequency	Stem	&	Leaf		
1.00	10	.	7		
2.00	11	.	03		
8.00	12	.	00233588		
8.00	13	.	00000558	Stem width:	10
11.00	14	.	00333355569	Each leaf:	1 case(s)
2.00	15	.	05		
3.00	16	.	005		
3.00	17	.	001		
1.00	18	.	5		
.00	19	.			
3.00	20	.	006		
2.00	Extremes		(210), (220)		

The stem and leaf display shows the pattern of data in the distribution while giving the reader the scores in the data set. Look at "stem" 14. It has 11 "leaves." Of the 44 weights, 11 were 140, 140, 143, 143, 143, 143, 145, 145, 145, 146, and 149.   Can you look at the display and determine what the other weights in the distribution are?

Critical Thinking Question: Rotate the stem and leaf display one–quarter turn (i.e., 90°) counterclockwise. From looking at the heights of each column, discern the shape of the distribution. Is the shape skewed?   Yes.   If skewed, is it positive or negative? Positive.

The frequency polygon allows you to "eyeball" the distribution to study the pattern the scores make. You graph Frequency on the *Y* axis and Height in inches on the *X* axis.

Look at the frequency column of the table for the 41 heights. What is the highest frequency that you see? It is 12 so the ordinate will need to be tall enough to go as high as 12. You could create an ordinate in increments of 2 (0, 2, 4, 6, 8, 10, 12), 3 (0, 3, 6, 9, 12) or 5 (0, 5, 10, 15).

On your abscissa, you want to represent your intervals using the interval midpoints. The midpoint is similar to the median. When the interval size is odd, the midpoint is an integer. When the interval size is even, the midpoint ends in .5. The interval 50–52 has an interval size of 3. What is its midpoint? 51. You then compute the midpoints for all of the intervals and write them on the abscissa. Remember to make the space on the abscissa between each midpoint equal.

After you draw your ordinate and abscissa, label the two axes, and write the increments for the measures, you are ready to start plotting the points. For each interval, put a dot above each midpoint that corresponds to the frequency in that interval. For the interval 50–52, you put a dot corresponding to a frequency of 1. When all of the dots are plotted, connect them with a straight edge. Your figure will look like:

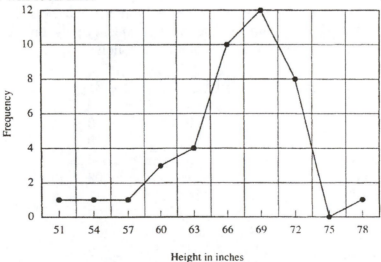

Height in inches

Study the figure. Is it a polygon? No. A polygon is an enclosed shape of many (i.e., poly) sides. The figure is currently not enclosed; it is open at both ends. To end up with a polygon, you add the midpoints of the intervals immediately before your smallest interval (48) and immediately after the largest interval (81) on the abscissa. Since both of these intervals will contain 0 scores, the dots for these intervals will be located on the abscissa. By connecting the open ends with these dots, you enclose the polygon. The finished frequency polygon appears below:

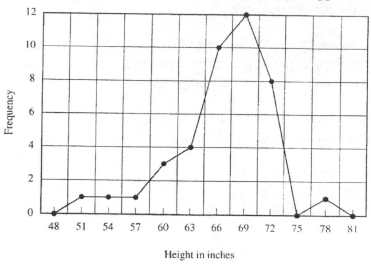

Height in inches

The frequency polygon is a picture of the data set! You can now *eyeball* the data. What do you see? Is the polygon's shape similar to the normal distribution? No. Do you see skewness? Not too much. Do you see any outliers? Yes, to the far right. What explanation could account for this outlier? In this case, the outlier was a member of the women's basketball team!

The Frequency Distribution Presenting Women's and Men's Heights Separately

It can be important to compare two groups' distributions with each other to gain insight about them. In the total class of 41 students, there are 28 women and 13 men. Can you draw the frequency polygons of the separate distributions for the women's and men's heights on the same graph? Yes, but slightly different from how the overall frequency polygon was drawn. First, create a frequency distribution table with five columns as follows:

Interval	Frequency$_{women}$	Frequency$_{men}$	RF$_{women}$	RF$_{men}$
77–79	1	0	.04	.00
74–76	0	0	.00	.00
71–73	1	7	.04	.00
68–70	7	5	.25	.54
65–67	9	1	.32	.08
62–64	4	0	.14	.31
59–61	3	0	.11	.08
56–58	1	0	.04	.00
53–55	1	0	.04	.00
50–52	1	0	.04	.00
	$\Sigma f = 28$	$\Sigma f = 13$	$\Sigma rf = 1.02$	$\Sigma rf = 1.01$

Next, calculate the mean, median, and mode for 28 women's heights and 13 men's heights.

For the women: $M = 64.89$, median $= 66$, mode $= 68$. For the men: $M = 69.77$, median $= 71$, mode $= 68$ and 72.

Now you are ready to graph the frequency polygons of both distributions on the same graph. Since the same intervals are used in both frequency distribution tables, the interval midpoints placed on the abscissa are the same. *However, instead of Frequency on the ordinate, the label will be Relative Frequency.* Why the difference? Because the two groups have unequal Ns, 28 women and 13 men. The unequal Ns make comparing the frequencies inappropriate. Converting the frequencies into relative frequencies solves this problem.

This graph differs in one other way. The two polygons are drawn with different lines. For example, the women's polygon might be a solid line, while the men's polygon is a dotted line. A legend informs the viewer which polygon is which. The two frequency polygons look as follows:

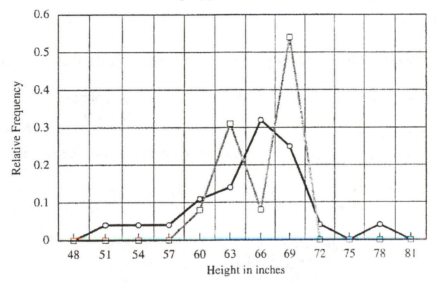
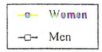

Computing the Weighted Mean

Suppose that you had the mean height for the 28 women and the mean height for the 13 men, and you wanted to know the overall mean for the total group of 41. How do you do this? Your first attempt might be to simply add the two means together and divide the sum by two. WRONG!

In statistics, the number of scores in the data set is of vital importance. For example, in the previous section, you had to use *relative frequencies* to label the ordinate of the graph on which the women's and men's polygons were drawn because of the unequal Ns of the two groups.

Why is the N so important? Generally speaking, the larger the sample size, the more similar the sample is to the population. You must take N into account when it differs between two samples.

When combining the means of two groups with unequal Ns to compute the overall mean, the formula to use is:

$$M_{overall} = \frac{(M_1 \times N_1) + (M_2 \times N_2)}{N_1 + N_2}$$

For the women's and men's groups, the formula looks like this:

$$M_{overall} = \frac{(64.89 \times 28) + (69.77 \times 13)}{28 + 13} = \frac{1816.92 + 907.01}{41} = 66.44 \text{ inches}$$

Problems

1. Indicate whether the following distributions are positively or negatively skewed:
   a. people's salaries
   b. housing costs
   c. test scores in a class where most students have done well but a few scored poorly

2. What is the optimal skewness of the distribution of your class' scores for the first statistics test?

3. The range of a particular distribution of college students' weights is 163 pounds.
   a. For your frequency distribution table, how many intervals will you have?
   b. What will the size of your intervals be?
   c. For making your frequency polygon, will your interval midpoints be integers or not?

4. The following data has an N of 15:   38, 39, 39, 40, 40, 40, 41, 41, 42, 44, 44, 45, 87, 87, 89
   a. What is the median?   b. Is the distribution skewed? If yes, is it positively or negatively?

5. In a positively skewed distribution, the median is _____ the mean.
      a. equal to   b. to the right of   c. to the left of   d. negative

6. If two–thirds of the sample have scores less than the mean, then the distribution is _____ (positively, negatively) skewed.

7. Sample is to population as _____ are to the even numbers less than 100.
   a. 1, 2, 90   b. 2, 4, 18, 92   c. 6, 28, 100   d. 96, 98   e. b, d   f. b, c, d   g. b, c

8. The preferred measure of central tendency for a skewed distribution is the
   a. mean   b. median   c. mode   d. a and b

For Problems 9–12, use the distribution 5, 9, 11, 11, 12, 14, 15, 17, 18, 98, 100:

9. What is the mean?   11. What is the mode?

10. What is the median?   12. Is the distribution positively or negatively skewed?

13. True or false: Social security number belongs to the ordinal scale of measurement.

Chapter 4

Measuring the CLUSTER or S P R E A D of the Data

To make a turkey sandwich, you spoon a dollop of mayonnaise on a piece of bread. Look at the shape of the mayonnaise—it's a lump. What happens to the height and width of the lump when you take a knife and start spreading the mayonnaise? As you spread the mayonnaise, the lump gets shorter and wider until you have covered the bread.

You have just mixed all of the ingredients to make cookie dough, and the mound of dough now sits on the kitchen counter. If you spread out the dough with a rolling pin, what happens to the height and width of the mound? The mound becomes shorter and wider. You finish rolling out the dough and then decide that you should knead it some more. You take the spread–out dough and push it together back into a mound. As you push the dough together, what happens to the height and width of the mound? It becomes taller and less wide because the dough has no place to go but up when you push it together.

Kurtosis

Numbers in a data set are like the mayonnaise and the dough. They can "form a mound" and cluster around the mean or be rolled out and spread away from the mean. When the data are clustered around the mean, the resulting frequency distribution looks like:

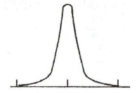

Look at this distribution. It is tall and narrow. Why is the distribution tall? What is making the distribution tall? Think of the lump of mayonnaise or mound of cookie dough. When they are tall and narrow, most of the mayonnaise or cookie dough is concentrated in the center. This distribution depicts the same arrangement of scores—most of the scores in the data set are clustered around the mean.

When the data are spread out from the mean, the resulting frequency distribution looks like:

Look at this distribution. It is short and wide. Why is the distribution short and wide? Because the data are spread out from the mean.

There is a word to describe the change in the height and width of a distribution—*kurtosis*. The tall and narrow distribution is called *leptokurtic*. The short and wide distribution is called *platykurtic*. The bell–shaped normal distribution has a *mesokurtic* shape.

The three prefixes are instructive. Lepto– means tall, thin, and peaked. Platy– means broad and flat as in the shape of the beak of the duckbilled platypus. You might have seen meso– used before.   The layer of skin *in between* the ectoderm and endoderm is called the mesoderm. Meso– means in the middle, in between, or intermediate.

Critical Thinking Question: Can you think of any other words that begin with meso? Mesopotamia was the land between the Tigris and Euphrates Rivers.

Critical Thinking Question: Look at the three frequency distribution tables below. All three tables contain heights collected from samples of 45 college students. Look at the intervals, the number of intervals, and the frequencies of each interval for the three tables. In Chapter 3, you graphed data from a frequency distribution table as a frequency polygon in order to "eyeball" the data. Imagine what the resulting polygons from each of the frequency distribution tables below would look like. Which of the three frequency distribution tables below correspond to a platykurtic, mesokurtic, or leptokurtic frequency polygon?

FREQUENCY DISTRIBUTION TABLE A		FREQUENCY DISTRIBUTION TABLE B		FREQUENCY DISTRIBUTION TABLE C	
Interval	Frequency	Interval	Frequency	Interval	Frequency
77-78	1	77-78	0	87-88	1
75-76	2	75-76	0	85-86	1
73-74	4	73-74	4	83-84	2
71-72	9	71-72	11	81-82	2
69-70	13	69-70	15	79-80	2
67-68	9	67-68	11	77-78	3
65-66	4	65-66	4	75-76	3
63-64	2	63-64	0	73-74	3
61-62	1	61-62	0	71-72	4
	45		45	69-70	4
				67-68	4
				65-66	3
				63-64	3
				61-62	3
				59-60	2
				57-58	2
				55-56	2
				53-54	1
					45

The three frequency polygons based on each of the frequency distribution tables appear on the next page. *Critical Thinking Question*: Which frequency distribution table matches with its corresponding polygon? Polygon A graphs Table B, Polygon B graphs Table A, and Polygon C graphs Table C.

FREQUENCY POLYGON A

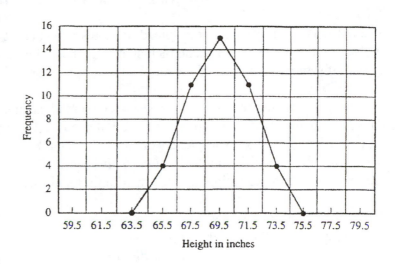

FREQUENCY POLYGON B

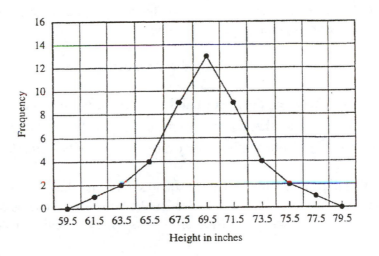

FREQUENCY POLYGON C

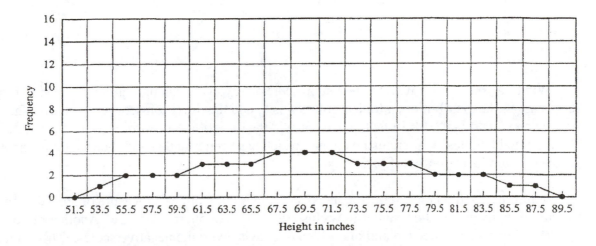

Measuring the Spread of the Data

The range is one measure of the spread of scores in a data set. The exclusive range is the highest score – the lowest score. The inclusive range is the highest score – the lowest score + 1. The spread of scores is different if the range of a sample of adult heights is 2" instead of 15".

In contrast to the range, the major measures of spread use the mean as the "anchor point" or referent of the distribution and define the spread of scores as the distance between the mean and each of the raw scores. For example, a weight of 165 pounds belongs to a distribution of adult weights whose mean is 155 pounds. *Critical Thinking Question*: What do you do to determine the difference between the weight and the distribution's mean? Subtract 155 from 165.

Just as you subtract the lowest score from the highest to find the range or distance between these two endpoints of the distribution, you subtract the mean from the raw score to find the distance between these two values. *The difference is the distance from the mean*, and this difference is the #1 reason for titling this book *Viva la Difference*! There are three measures for the spread of the scores in a data set based on the distance between the mean and the raw scores: sum of squares, variance, and standard deviation.

The Sum of Squares Computed Using the Deviation Scores

The difference between the mean and the raw score is called the deviation score and is symbolized as $X - M$ (note that the sample mean can also be symbolized as $\overline{X}$ and the deviation score is then $X - \overline{X}$). Your textbook may also symbolize the deviation score as x.

The first step to calculate the sum of squares is to find the deviation scores for all of the raw scores in the distribution. A sample of five children contains the ages of 3, 6, 7, 9, and 15 years. What is the mean of this sample? 8 years. Here is what the deviation scores are:

raw scores	$X - M$		deviation scores
15	$15 - 8$	=	7
9	$9 - 8$	=	1
7	$7 - 8$	=	-1
6	$6 - 8$	=	-2
3	$3 - 8$	=	-5

What do you now do with the deviation scores? If you are calculating how spread out the scores are in the distribution, a logical solution is to add the deviation scores together. What is the sum when you do that? 0! In fact, when you add the deviation scores for any distribution, they always equal 0, unless you are using a mean which has been rounded and then rounding error may produce a sum just slightly different from 0.

Summing the deviation scores in any data set will not work as a measure of spread because the sum will always be 0 no matter how clustered or spread out the data are. Another option for the deviation scores is needed and one which does away with the negative scores. This option is squaring the deviation scores. The *squares* in the term *sum of squares* refers to the squaring of the deviation scores. So now you have:

raw scores	$X - M$		deviation scores	deviation scores^2
15	$15 - 8$	$=$	7	49
9	$9 - 8$	$=$	1	1
7	$7 - 8$	$=$	-1	1
6	$6 - 8$	$=$	-2	4
3	$3 - 8$	$=$	-5	25

When you sum the squared deviation scores, you get 80. 80 is the sum of squares. *Critical thinking question*: 80 what? What are the units for the 80? The units are *years*^{2} . Understanding years^2 makes understanding the sum of squares difficult.

The sum of squares formula using the deviation scores is called the *deviational formula* and can be written several ways. For example, what was just done is written as *sum of squares* = $SS = \Sigma(X - M)^2 = \Sigma x^2$. If the sample mean in your textbook is symbolized as $\overline{X}$, then the deviational formula is written as sum of squares = $SS = \Sigma(X = \overline{X})^2$

CAUTION: You do not want to treat the Σ as a variable and incorrectly compute the sum of squares as $(\Sigma X - M)^2$ or $\Sigma X^2 - M^2$ or $\Sigma X - M^2$. *These formulas are incorrect.* Remember that Σ symbolizes the verb *add* and does not represent a variable.

The Sum of Squares Computed Using the Raw Scores

Critical Thinking Question: Which is larger, the raw scores or the deviation scores?   The raw scores. *Critical Thinking Question*: Which is larger, ΣX^2 or Σx^2 ? Because the raw scores are larger than the deviation scores, then ΣX^2 will be larger than Σx^2, which is the sum of squares.

However, the sum of squares can be computed using the raw scores instead of the deviation scores. When the sum of squares is computed using the raw scores, the formula starts with ΣX^2.

Critical Thinking Question: Because the raw score formula for the sum of squares starts with ΣX^2 and because ΣX^2 *is larger than* Σx^2, do you add to or subtract from ΣX^2 to compute the sum of squares? We have to make ΣX^2 smaller so something has to be removed or subtracted.

What is being removed from ΣX^2 is $\dfrac{(\Sigma X)^2}{N}$. The formula for computing the sum of squares using the raw scores instead of the deviation scores is called the *computational formula for the sum of squares*:

$$\text{sum of squares} = \Sigma X^2 - \frac{(\Sigma X)^2}{N}$$

Some textbooks label the $\dfrac{(\Sigma X)^2}{N}$ expression as C, which stands for *correction factor*.

Using the same five children's ages from the above table, you now have

Raw scores	Raw scores^2
15	225
9	81
7	49
6	36
3	9
40	400

sum of squares $= \Sigma X^2 - \dfrac{(\Sigma X)^2}{N} =$

$400 - \dfrac{40^2}{4} = 400 - 320 = 80$

CAUTION: ΣX^2 and $(\Sigma X)^2$ are not the same. ΣX^2 is the raw scores *squared then summed*, whereas $(\Sigma X)^2$ is the raw scores *summed then squared*. They do not equal each other.

The Variance

The variance is the average of the squared deviation scores. To find the average raw score (i.e., the mean), you divide the sum of the raw scores (ΣX) by N. You follow the same thinking to find the variance. You divide the sum of the squared deviation scores by N.

$$\text{variance} = V = \frac{SS}{N} = \frac{\Sigma x^2}{N} = \frac{\Sigma(X - M)^2}{N} = \frac{\Sigma X^2 - \dfrac{(\Sigma X)^2}{N}}{N}$$

For the data set of five children's ages, the SS is 80 so the variance is $\dfrac{80}{5} = 16\, years^2$. What is a *year^2*? Just like the sum of squares, understanding the variance is more complicated because of the squaring of the unit of measurement.

The Standard Deviation

The standard deviation is the square root of the variance. Thus,

$$SD = \sqrt{\text{variance}} = \sqrt{\frac{SS}{N}} = \sqrt{\frac{\Sigma x^2}{N}} = \sqrt{\frac{\Sigma(X - M)^2}{N}} = \sqrt{\frac{\Sigma X^2 - \dfrac{(\Sigma X)^2}{N}}{N}}$$

Computationally, dividing the sum of squares by N when you compute the variance counteracts summing the squared deviation scores. Then taking the square root of the variance when you compute the standard deviation counteracts squaring the deviation scores. For the data set of five children's ages, the standard deviation is the $\sqrt{\text{variance}} = \sqrt{16\,\text{years}^2} = 4\,\text{years}$. Note that the unit of measurement is now years rather than years^2 making the standard deviation the most understandable of the three measures of variability.

Of the three measures of variability, the standard deviation is the one presented in research articles because the units of the standard deviation are the same as the units for the raw scores and the mean—years, pounds, inches, number of words remembered in a list, points on a test. In the next chapter, you will see the standard deviation used in a very important context.

Variability Example

Here is a work sheet for a data set of 13 scores from the first examination of a section of introductory statistics I taught several years ago:

X	X^2	$(X - M)$ or x	x^2
97	9409	14.46	209.09
96	9216	13.46	181.17
94	8836	11.46	131.33
92	8464	9.46	89.49
89	7921	6.46	41.73
88	7744	5.46	29.81
87	7569	4.46	19.89
83	6889	.46	.21
77	5929	−5.54	30.69
73	5329	−9.54	91.01
72	5184	−10.54	111.09
69	4761	−13.54	183.33
+ 56	+3136	−26.54	+ 704.37
1073	90387	−.02	1823.32

$\Sigma X \qquad\qquad \Sigma X^2 \qquad\qquad\qquad \Sigma x \qquad \Sigma x^2 = \Sigma(X - M)^2$

range = 42   mean = M = 82.54

sum of squares deviational formula $= \Sigma x^2 = \Sigma(X - M)^2 = 1823.32$

sum of squares computational formula $= \Sigma X^2 - \dfrac{(\Sigma X)^2}{N} = 90{,}387 - \dfrac{(1073)^2}{13} = 1823.32$

variance $= \dfrac{\text{sum of squares}}{N} = \sqrt{\dfrac{1823.23}{13}} = 140.25$

$$\text{standard deviation} = \sqrt{\frac{\text{sum of squares}}{N}} = \sqrt{\frac{1823.23}{13}} = \sqrt{140.25} = 11.84$$

The Words Behind the Numbers

The standard deviation measures how clustered or spread out the scores are from the mean of the data set. For example, the smaller the distribution's standard deviation becomes, the a) more clustered (i.e., less spread out) or closer the data are around the mean, b) more homogeneous is the data set, and c) more leptokurtic is the shape of the distribution. Conversely, the larger the distribution's standard deviation becomes, then the a) more spread out the data are from the mean, b) more heterogeneous is the data set, and c) more platykurtic is the shape of the distribution.

CAUTION: Be careful comparing standard deviations across distributions of *different measurements*. For example, the standard deviation for a leptokurtic distribution of weights could be greater than the standard deviation for a platykurtic distribution of heights.

Perspectives

Chapter 1 presented learning statistics from two perspectives. From the *doer of research perspective*, you collect the data and compute the descriptive statistics about the data set's variability. You compute the sum of squares, variance, and standard deviation. From the *consumer of research perspective*, can you compute the variance and sum of squares when you read the standard deviation reported in a research article? Yes: $V = SD^2$ and $SS = V \cdot N$. If an author reports the Ns, means and standard deviations in a research article, you can compute the variance, sum of squares, and ΣX (remember that $\Sigma X = M \cdot N$) for the groups.

Application

On the next page is the *Birth to 36 months: Girls Length-for-age and Weight-for-age percentiles Growth Charts* (there are also charts for boys) available from the Centers for Disease Control at http://www.cdc.gov/growthcharts/data/set1clinical/cj41c018.pdf. The growth charts begin at birth with increments every three months through 36 months. The Weight chart presents pounds or kilograms. Look at the Weight chart, and you will see 7 growth curves with a boldface curve in the middle flanked on either side by two lightface curves and an end boldface curve. The middle curve represents the mean, median, and mode so you can see the spread of weights in the distribution for any age. The growth chart uses variability as a tool for health. Very low birth is implicated in in physical and cognitive problems including schizophrenia. Extreme weights at any age alert pediatricians to check for poor nutrition or an endocrine disorder.

(From K. A. Weaver (1999). The statistically marvelous medical growth chart: A tool for teaching variability. *Teaching of Psychology, 26*, 284-286. Used by permission, Lawrence Erlbaum.)

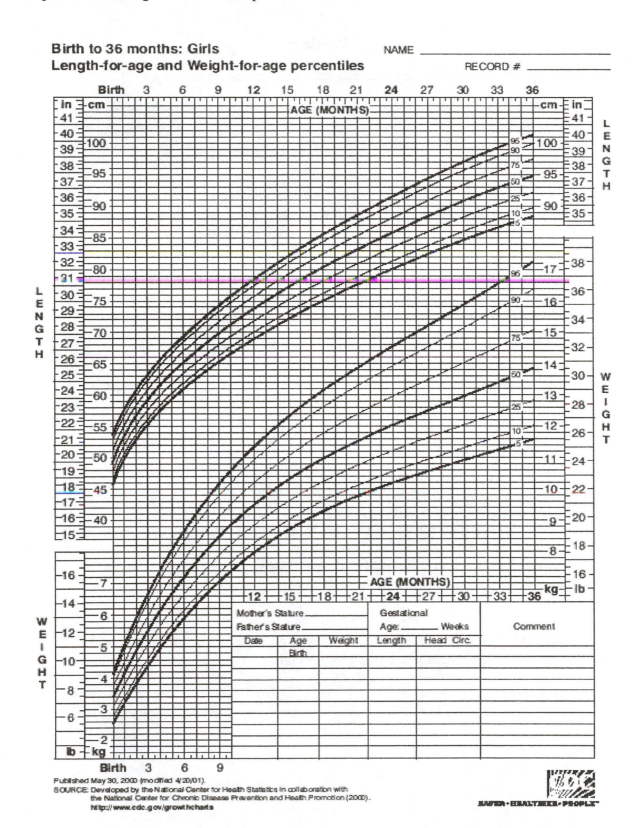

Problems

1. What is the standard deviation of the data set whose scores are all the same?

2. True or false: Standard deviation, variance, and sum of squares can only be positive.

3. The standard deviation is to measures of _____ as the mean is to measures of central tendency.

4. In a research journal in your campus library's Periodical Section, find a table of means and standard deviations in an article. Can you compute the variance and sum of squares?

5. The sum of the squared deviation scores is to the variance as the sum of the raw scores is to the _____.

6. Homogeneity is to heterogeneity as _____ is to spread.

   For Problems 7 through 11, use the data set 75, 80, 85, 90, 95.

7. What is the mean of this data set?

8. What is the sum of squares of this data set?

9. What is the variance of this data set?

10. What is the standard deviation of this data set?

11. Which of these distributions is more homogeneous: 3, 5, 6, 7, 8   OR   3, 9, 12, 15, 18? Compute the standard deviations of the two distributions to support your answer.

12. On the first exam, Section A has a $M = 83$ and $SD = 6$ and Section B has a $M = 88$ and $SD = 8$. What are the variances of the two distributions, and which distribution is more platykurtic?

13. Find out your birth weight and locate it on the growth chart on the preceding page. How variable is your birth weight to the mean birth weight?

Chapter 5

Locating the Data

The normal curve is actually a theoretical distribution produced from the equation:

$$f(X) = \frac{1}{\sigma\sqrt{2\pi}} e^{-\frac{(X-\mu)^2}{2\sigma^2}}$$

The symbols μ and σ stand for the population mean and population standard deviation, respectively. Also, the $f(X)$ means "function of X," which is another way of thinking about the measurement plotted on the Y axis. $f(X)$ is the Y measurement because the value being graphed on the Y axis is a function of or dependent upon the corresponding value being graphed on the X axis. In Chapter 9, the terms independent variable and dependent variable will be further elaborated on p. 67.

Critical Thinking Question: Given what you have read in this preceding paragraph, on which axis do you expect to find the dependent variable graphed and on which axis do you expect to find the independent variable graphed? The dependent variable is graphed on the Y axis and the independent variable is graphed on the X axis.

Trivia question: Although the formula for the normal curve is found in most statistics textbooks, where did I get the above formula from? I copied it from one side of Germany's 10 mark bill. The formula appears with a picture of the normal curve in the background and the image of Karl Friedrich Gauss (1777–1855), who is the mathematicians credited with discovering the formula.

There actually is a family of normal curves because μ and σ can have different values. These curves differ in terms of their kurtosis which is determined by the population's standard deviation or σ. However, when μ and σ are 0 and 1, respectively, the normal curve becomes the *standard normal curve*. There is only one standard normal curve.

The Standard Normal Curve

The standard normal curve is a frequency distribution. Frequency is graphed on the ordinate; the scores are graphed on the abscissa. However, the raw scores can be graphed according to their *z* score equivalents. Examine this standard normal curve:

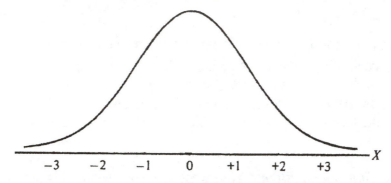

Look at the values on the abscissa—why are these values used and not the raw scores. Because they represent *z* scores. What are *z* scores? They *ARE* the raw scores but expressed in units of the standard deviation rather than units of inches or pounds or points on a test. A *z* score of 0 equals the mean. Negative *z* scores are located to the left of the mean (remember the number line on p. 6 in Chapter 1) and correspond to those raw scores which are smaller than the mean. Positive *z* scores are the raw scores greater than and thus to the right of the mean.

The standard normal curve consists of a central region of convexness bordered by two outer areas of concaveness. Where the convexness becomes concave is approximately where the *z* scores of ±1 standard deviations are located. The same distance along the abscissa between the mean and 1 standard deviation exists between 1 and 2 standard deviations and between 2 and 3 standard deviations.

Areas Under the Standard Normal Curve

Examine the standard normal curve below, now with its area divided into sections:

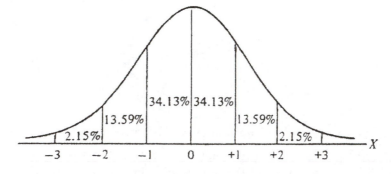

The percent given in each section of the curve is the percent of the total area occupied by that section. Notice that the boundaries of each chunk are –3, –2, –1, 0, 1, 2, and 3. It is possible, however, to find the percent of the total area of any of the sections under the normal curve, even if the boundaries are not integers (e.g., –1.35 or .89). To do this, you can access the Standard Normal *z* Distribution Table at http://www.statsoft.com/textbook/distribution-tables/#z or use the *z* Distribution Table in the back of a statistics textbook.

Knowing the areas of sections of the normal curve is valuable because *the percent of the total area in a particular section corresponds to the percent of the scores in the distribution which belong in that section.* *Critical Thinking Question*: If 50% of the total area under the normal curve is above the mean, then how many scores in the normal distribution are above the mean? 50%.

In the *z* Distribution Table, look at the *z* score of 1.00, and you will find either 34.13% or .3413. This means that 34.13% of the total area in a normal distribution is occupied by that section of the distribution bordered by the mean and 1 standard deviation from the mean. Thus, 34.13% of the scores in the data set have values located in this section. If the distribution's *N* is 10,000 scores, then 3,413 scores (i.e., 10,000 x .3413) have values between the mean and 1 standard deviation.

Converting Raw Scores to *z* Scores

Who is taller—someone who is 6 feet tall or someone who is 2 yards tall? What a silly question—6 feet and 2 yards are the same heights. But does 6 = 2? Of course not, so the units of measurement of feet and yards are important. How do you convert feet to yards? There is a formula for the conversion; the number of feet divided by 3 equals the number of yards.

Like converting feet to yards, converting a raw score to a *z* score requires a formula and paying attention to the units of measurement. The formula is:

$$z = \frac{X - M}{SD}$$

In the last chapter, for a data set of five children's ages—3, 6, 7, 9, and 15—you computed the sum of squares, variance, and standard deviation. The mean was 8, and the standard deviation was 4. Converting the raw score of 15 to its *z* score equivalent looks like this:

$$z = \frac{X - M}{SD} = \frac{15 - 8}{4} = \frac{7}{4} = 1.75$$

Does 15 = 1.75? No, so you must take into account the units of measurement in order to make sense of the conversion. Thus, 15 years = 1.75 standard deviations. But what does this mean? *The advantage of converting a raw score into its standard deviation equivalent is that you can locate where the raw score is on the normal curve relative to the other raw scores in the distribution.* The age of 15 years is located 1.75 standard deviations to the right of the mean.

Compute the *z* scores for the other four ages and complete the table below:

Raw scores	z
15	1.75
9	
7	
6	
3	

Now locate the five z scores on the standard normal curve to practice locating these raw scores.

z Scores at Work

Example 1. You receive a grade of 90 on a test. Wow, 90! What a fine grade! What do you conclude about the 90 if the 90 converts to a raw score of −1.75? That 90 is actually a low score when compared with the other scores in the distribution.

Example 2. You receive a score of 70 on a test, and the instructor tells you that the 70 corresponds to a z score of 2.15. How should you feel about your score? Your score may be the highest in the class!

z scores tell us where in the distribution the raw score is located relative to every other score in the distribution. The raw score itself cannot communicate this information about location.

Converting the z Score to the Raw Score

You may need to convert a z score to a raw score. This conversion can be done by algebraically rearranging the z score formula so that X is isolated on one side of the equal sign.

Critical Thinking Question: Can you start with the z score formula and make the necessary steps to isolate X on one side of the equal sign? You will end up with the following equation:

$$X = (z)(SD) + M$$

z Scores and Percentile Scores

The area under the normal curve can be divided into 100 segments. Each segment contains 1% of the area and is called a percentile.

CAUTION: Do not regard these segments as having equal widths (like slices of bread in a loaf) because they do not. In the tails of the distribution where the curve is close to the abscissa, each segment is shorter and thus must be wider than the segments in the center of the distribution where the curve is taller to make up 1% of the area.

The percentile acts the same way as the z score; it locates the score relative to the other scores in the distribution. Thus, a score located at the 50th percentile in a normal distribution would be located at the mean (where the z score is 0) with one-half of the scores smaller and the other one-half of the scores larger. A score located at the 16th percentile in the distribution means that 16%

of the scores in the distribution are less than that score while 84% of the scores in the distribution are greater than that score.

Critical Thinking Question: Can you figure out what *z* score corresponds to the 16th percentile? It would be a *z* score of -1. Conversely, the 84th percentile is located at a *z* score of +1.

Look at the right side of the growth chart for weight on p. 37 and find 5, 10, 25, 50, 75, 90, and 95 that label the seven curves. These numbers are percentiles in the distribution. A weight located in the 75th percentile of the distribution means a weight that is greater than 75% of the weights but less than 25% percent of the remaining weights. Note that you can translate percentiles into *z* scores using the table at http://images.tutorvista.com/cms/images/67/Z-Score-Percentile-Table.jpg.

Applying Variability and Location[1]

The medical growth charts used by pediatricians and local health departments are based on locating a child's weight, length, or head circumference relative to the other children in the norm groups. The charts were last revised in 2000 by the National Center for Health Statistics (NCHS), which is part of the Centers for Disease Control, to provide a more accurate instrument for evaluating the growth status of infants, children, and adolescents in the United States.

Look at the distribution of weights from birth through three years of age in the growth chart on page 37. Age in three month increments is graphed along the *X* axis and along the *Y* axis is weight in both kilograms and pounds. *Critical Thinking Question:* Why is age graphed on the abscissa and weight graphed on the ordinate? If $Y = f(X)$, then *Y* is a function of or dependent upon *X* for its value. Thus, weight is a function of or dependent upon the age of the child for its value.

Critical Thinking Question: Why include kilograms? So that the chart can be used wherever weight is measured in kilograms.

Study the pattern on the growth chart. See how the range of weights is small at birth and then expands across the three years. Do you know your birth weight? Find out what your birth weight was and plot it on the chart. Convert your birth weight to percentile to see where you are located on the distribution of weights with all of the other newborns.

Look at the 5th, 10th, 25th, 50th, 75th, 90th, and 95th percentiles marked on the upper right half of the chart. The mean is the 50 percentile and is indicated by the bold line.

Visit http://blogs.wsj.com/health/2011/07/11/whats-the-best-size-for-a-newborn-baby/ and read the article about this child. Look to see where he would be placed on the boy's weight chart

---

[1]This section is excerpted from Weaver, K. A. (1999). The statistically marvelous medical growth chart: Tool for teaching variability. *Teaching of Psychology, 26,* 284-286. Used by permission, Lawrence Erlbaum Associates.

(http://www.cdc.gov/growthcharts/data/set1clinical/cj41c017.pdf) for both 16 months and 36 months. At the 99th percentile, the toddler's weight is an extreme outlier having a very low frequency of occurrence. Extreme variability of this sort can indicate a possible problem that may lead a pediatrician to order follow-up tests on growth hormones, nutrition, or intellectual development. Other articles about this child report that a pediatric endocrinologist found no evidence of a glandular disorder. Why would this finding be relevant for this child?

What adjectives would you use to describe this toddler's weight? You might think of rare, extreme, strange, and very unusual. Do you recognize that these terms are statistical in nature and are directly applicable to variability? When you use these terms, you are indicating that you have a sense of a) what is average and b) how different what you have encountered is from that average.

Birth weight is one important indicator of child development. Child psychology and developmental psychology courses cover fetal and infant growth and often address the physical and cognitive problems of very low birth weight, low birth weight, and small-for-gestational-age infants (e.g., see Feldman, 2011). The abnormal psychology course may also include reference to the association between low birth weight and schizophrenia (Larsen, J. K., Bendsen, B. B, Foldager, L., & Munk-Jørgensen, 2010; abstract at http://www.ncbi.nlm.nih.gov/pubmed/25385215).

Problems

1. True or false: Deviation score, z score, variance, sum of squares, and standard deviation can have positive and negative values

For Problems 2 through 10, assume a normal distribution of a sample with $N = 200$, $M = 21$, and $SD = 3$. Use the z Distribution Table at http://www.statsoft.com/textbook/distribution-tables/#z or at the back of a textbook to solve these problems.

2. Convert the raw score of 17 to a z score.

3. What percent of the total area resides to the left of 17?

4. What percent of the total area resides above 17?

5. How many scores in the data set are greater than 21?

6. How many scores in the data set are greater than 17?

7. How many scores in the data set are between 17 and 21?

8. What is the percent of the total area between 21 and 23?

9. What is 20 expressed as a z score?

10. Which raw score is located farther from the mean: 15 or 26?

Chapter 6

Probability

What are inferential statistics and how is probability connected to them? Introducing probability is a cue that the course's focus is changing from describing a data set to using sample data to infer about the population for which you do not know the parameters. You estimate the population mean and population standard deviation by generalizing from the few (the sample) to the many (the population). These estimates answer a variety of statistical questions.

But first, you need to determine the accuracy of your estimates. Sample size and whether the sample has been randomly selected or not from the population are two factors. Another factor is probability, which informs you of the likelihood that the estimate of the parameter you have produced based on the sample actually is a good estimate of the population parameter being studied.

Values for Probability

The values for probability range from 0 to 1. A probability of 0 means that a future event will not happen. A probability of 1 means that a future event is certain to happen. All values of probability are positive.

As evidenced by the number of English expressions and words referring to probability, we are quite comfortable with understanding the concept. Consider this list and write down what you think the probability value is for each of them:

it's a done deal	maybe	count on it
never happ'n, Cap'n	could be	might be
it's a sure thing	possibly	should be
take it to the bank	probably	ought to be

Computing Probability

One way of computing probability is to divide the number of events which satisfy some criterion by the total number of events. The resulting proportion is the probability.

Example 1. The criterion is to select a red card from a deck of cards. There are 26 red cards which satisfy the criterion and 52 possible cards to select, so the probability of selecting a red card is $\frac{26}{52} = .50$.

Example 2. The criterion is to pick a horse to win the race when eight horses of equal ability are racing. One horse satisfies the criterion and eight possible horses can win, so the probability of selecting the correct horse is $\frac{1}{8} = .125 = .13$.

Example 3. The criterion is to roll a die and have the three dots facing up. The die has six sides, and only one satisfies the criterion, so the probability of rolling a 3 is $\frac{1}{6} = .16\bar{6} = .17$.

Combining Probabilities

Events can occur in two different combinations. The first combination is that more than one event satisfies the criterion. Let's restate Example 3 above to be: The criterion is to roll a die and have either the three dots *OR* the six dots facing up. *Critical Thinking Question*: Is rolling a 3 or 6 more or less probable than rolling a 3? Rolling a 3 or 6 is more probable than rolling just a 3.

Finding this probability requires using the "add–or" rule: There are six sides to the die and now two of them satisfy the criterion. The probability of rolling a 3 or 6 is $\frac{1}{6} + \frac{1}{6} = \frac{2}{6} = \frac{1}{3} = .33$.

For the second combination, the events can occur in sequence. Again using Example 3: The criteria is to roll the die once and have the three dots facing up *AND* then roll the die again and have the six dots facing up. *Critical Thinking Question*: Is rolling a 3 and then rolling a 6 more or less probable than rolling a 3? Rolling a 3 and then rolling a 6 is less probable than rolling a 3.

Finding the probability requires using the "mult–and" rule: The probability of rolling a 3 and then rolling a 6 is $\frac{1}{6} \times \frac{1}{6} = \frac{1}{36} = .02\overline{7} = .03$.

Perhaps you have seen on television the process of selecting the winning lottery number. If the winning number has 5 digits, each digit can be one of 10 values (i.e., 0 through 9). The probability of selecting the winning number is then $\frac{1}{10} \times \frac{1}{10} \times \frac{1}{10} \times \frac{1}{10} \times \frac{1}{10} = \frac{1}{100,000} = .00001$.

With a probability of success so small, why are people willing to buy a lottery ticket? Because the potential winning is worth the risk of losing a dollar or two.

Probability and the Normal Curve

Areas under the normal curve can also be used as the probabilities of selecting scores from a normal distribution. For example, what is the likelihood of selecting a score from a normal distribution whose value is greater than the mean? .50.

Critical Thinking Question: What is the probability of selecting a score from a normal distribution whose value is less than a *z* score of –1? .16.

The areas in the tails of the normal distribution are small. Consequently, the probability of selecting an unusual score located in one of the tails is small. We expect the probability of randomly selecting an unusual score to be small because these extreme scores occur rarely, and a rare event has a small probability of occurrence.

What is the probability of selecting a score whose *z* score is either less than –1.96 or greater than 1.96? The answer is .025 + .025 = .05. Thus, you randomly select scores from the extreme tails of the normal distribution 5 times out of 100 on average.

Three Different Types of Probability Problems

Below are three different types of problems using probability. Note that the third problem is one which will require content from the *z* test presented in the next chapter.

1. What is the probability of selecting a score greater than 90 from a normal distribution of scores whose mean is 85 and standard deviation is 5?

$$M = 85 \qquad SD = 5 \qquad X = 90 \qquad z = \frac{X - M}{SD} = \frac{90 - 85}{5} = \frac{5}{5} = 1$$

From the table of areas under the normal curve in the back of your statistics textbook, the probability of selecting a score between 85 and 90 is thus .3413. Consequently, the probability of selecting a score greater than 90 is .5000 – .3413 = .1587, which is rounded to .16.

2. What is the probability of selecting a score from the most extreme 5% of the scores in a normal distribution and what *z* scores form the boundaries to these areas?

The probability of selecting a score from the most extreme 5% of a normal distribution is .05.

The normal distribution has two tails so the most extreme 5% consists of the most extreme 2.50% in the left tail and the most extreme 2.50% in the right tail. Thus, in one of the halves of the distribution, the portion of the area not part of the extreme section is 50% – 2.50% or 47.50% or .4750. Look at http://www.statsoft.com/textbook/distribution-tables/#z or the table in the back of your textbook containing areas under the normal curve. Find the areas between the mean and the *z* score and then look for 47.50% or .4750. What *z* score does 47.50% or .4750 correspond to? It is 1.96. Because half of the extreme 5% is located in the left tail and the other half is located in the right tail, the boundaries into these areas are located at ±1.96.

3. What is the probability of randomly selecting from a population of test scores whose mean is 80 and standard deviation is 14, a sample of 49 test scores whose mean is less than 75? Another way of asking this question is what is the probability of randomly selecting from the sampling distribution of the mean, whose mean is 80 and standard error of the mean is 2, a sample of 49 whose mean is less than 75?

$$\mu = 80 \qquad \sigma = 14 \quad N = 49 \quad M = 75$$

You first find the standard error of the mean: $\sigma_M = \dfrac{\sigma}{\sqrt{N}} = \dfrac{14}{\sqrt{49}} = \dfrac{14}{7} = 2$

Then plug σ_M into the *z* test formula: $z = \dfrac{M - \mu}{\sigma_M} = \dfrac{75 - 80}{2} = \dfrac{-5}{2} = -2.50$

From the table at http://www.statsoft.com/textbook/distribution-tables/#z or in the back of your statistics textbook, the probability of selecting a sample of 49 whose mean ranges from 75 to 80 is .4938. Thus, the probability of selecting a sample whose mean is less than 75 is between .50 – .4938 = .0062, which is rounded to .01. We would classify the probability of this event as being rare.

Inclusion and Exclusion: Scenario 1

The z score is the location of the raw score relative to all of the other raw scores in the distribution. The probability of randomly selecting a score from a distribution becomes smaller and smaller as the score is located farther and farther from the mean. This makes sense as you look at the shape of the normal distribution because most scores are located around the mean, and the number of scores located in the tails is very small.

Suppose that you only had access to a distribution's mean and standard deviation but not to any of the scores making up the distribution. Now suppose that someone gives you a score and asks you to make a judgment as to whether or not the score belongs to this distribution. Good grief! How are you going to accomplish this task? The answer is that you are going to base your judgment on probability and logic. Here is how that would work:

Using the distribution's mean and standard deviation and the z score formula, you convert the score into a z score to determine where the score is located on the distribution. Based on this location, you determine some probabilities of selecting that score from that distribution. For example, there is a probability of .68 of selecting a score located between −1 and +1 standard deviations. If your raw score is located in this area, then there is a strong probability that the score belongs to this distribution, and you respond yes to *including* the score as belonging to the data set.

Suppose, however, that your score is located in one of the extreme tails of the distribution. The probability of this score being part of the distribution is now small. Logically, you reason that the probability of that extreme score belonging to the distribution is so small that it is more likely that the score comes from some other distribution. You say no about this score belonging, *excluding* it as belonging to this distribution. Now, even though you are using probability to base your decision to exclude the score, this decision could be wrong. Statistically speaking, the probability of being wrong must be kept small and is expressed as $p < .05$ or $p < .01$. This probability of making the erroneous decision to exclude the score when in fact it is part of the data set is called alpha and is explained more in the next chapter

Inclusion and Exclusion: Scenario 2[2]

Imagine a wind-sheltered orchard of trees during the autumn. As the trees shed their leaves, the leaves fall to the ground forming piles around the tree trunks. Each pile is tallest at the tree trunk and decreases in height farther away from the tree. The shape of these piles is similar to the outline of the standard normal distribution.

You drive to the orchard and see the piles surrounding the trees. You walk up to a tree, and standing by its trunk, pick a leaf from the pile. You are asked to make a judgment as to whether the leaf has come from this tree or some other tree in the orchard. What is the probability that the leaf you are holding fell from the tree under which you are standing? Very high, almost certainty. You say yes to this leaf coming from this tree and including this leaf in the pile of leaves from this tree.

---

[2]Excerpted from Weaver, K. A. (1992). Elaborating selected statistical concepts with common experience. *Teaching of Psychology, 19*, 178-179. Used by permission, Lawrence Erlbaum Associates.

Now you walk 60 feet away from the tree trunk and pick up a leaf from the edge of the pile. What is the probability that this leaf has come from that same tree. How sure are you that this leaf has come from the same tree and not a neighboring one?

As you walk farther away from the tree, a point is ultimately reached beyond which you say no to any leaf coming from the tree. You would exclude any of these leaves from coming from the tree because *there would be a greater probability of these leaves coming from a neighboring tree.*

In the next several chapters, the logic and decision making described in these two scenarios form the basis for null hypothesis significance testing.

Randomness

Randomness is defined based on probability. Members of a sample are randomly selected from a population if all members of the population have an equal probability of being selected. How can this equal probability be attained? One way is to write down each member of the population on a separate slip of paper, all slips being the same size and type of paper. The slips are then put into a box, mixed, and blindly selected one at a time.

Another technique for random selection is to use a random number table (see Table 6-1 on the next page) to assign a number to each member of the group. Then the members with the highest or lowest or even or odd (your choice) assigned numbers constitute the sample.

Problems

For Problems 1 and 2: The mean and standard deviation of a normal distribution are 100 and 15.

1. What is the probability of selecting a score which is less than 85?

2. What raw scores form the boundary to the most extreme 5% of the distribution's total area?

3. True or false: Standard deviation, sum of squares, and probability can never be negative.

4. Look for the expression $p < .05$ in the results sections of research articles in your favorite journal. Explain to yourself what you think it means.

5. The extreme 10% of the area under the normal curve is bounded by what $\pm z$ scores?

6. What is the probability of selecting on one draw either an ace or a three?

7. What is the probability of flipping a coin and getting four heads in a row?

8. True or false: Random sampling means that every member of the population has an equal probability of being selected.

9. True of false: The smallest value for probability equals the value of the standard deviation when all the scores in the distribution are the same.

10. Use the random number table on the facing page to create a sample from a group (e.g., your class, sorority, fraternity, residence hall floor, study group).

Table 6-1. Random Number Table

007	662	427	628	466	232	199	713	389	615	795	114	066
564	187	437	898	282	317	405	737	478	595	617	497	214
509	745	569	814	915	037	409	117	852	349	471	753	883
511	742	290	297	026	529	565	276	858	853	582	837	228
226	347	563	681	627	311	598	918	016	678	477	414	761
815	417	072	752	316	647	458	914	256	262	420	479	192
010	452	364	572	248	415	295	562	483	726	408	586	732
424	820	259	440	009	094	147	832	462	860	496	308	743
504	261	872	023	829	546	892	791	552	243	631	184	178
913	516	644	368	407	090	894	495	154	237	683	312	272
413	348	024	749	656	071	435	730	800	296	423	305	802
700	307	315	422	649	105	284	171	438	670	387	602	782
213	902	756	880	792	607	189	014	600	077	222	708	485
100	131	148	294	903	398	544	584	783	135	794	208	019
557	058	886	787	130	464	300	334	878	784	692	762	170
481	028	531	084	456	606	591	309	215	166	664	519	363
343	561	917	808	043	548	790	725	380	402	224	395	515
774	722	721	210	788	211	235	040	612	283	017	045	168
359	344	635	370	710	819	797	108	225	657	200	758	558
623	378	202	693	545	778	379	008	003	487	101	863	030
158	863	385	250	680	269	709	807	273	009	257	640	523
909	055	354	144	810	767	901	554	120	093	123	537	877
322	221	651	337	724	164	454	442	012	000	861	448	374
430	291	059	575	242	073	550	870	518	621	054	685	216
025	703	142	041	075	397	302	411	676	065	125	840	622
521	851	258	031	912	103	592	875	589	532	085	321	118
083	842	919	541	209	690	547	246	039	107	350	082	234
465	416	160	132	335	679	355	825	155	744	099	070	369
691	597	021	799	855	145	754	712	711	111	798	157	540
868	313	474	446	763	299	539	492	394	038	848	839	074
854	163	266	095	116	428	556	306	587	658	543	817	605
641	190	325	384	831	702	696	029	746	382	704	247	365
809	857	601	916	895	716	286	404	524	109	319	092	436
450	188	418	460	241	352	672	426	133	728	080	326	776
014	573	910	212	229	823	433	603	490	372	036	124	699
659	472	050	489	088	104	223	613	503	896	906	434	805
254	205	177	772	001	336	773	169	859	849	821	826	180
106	179	866	512	620	882	695	596	467	429	771	265	274
614	191	701	122	279	338	252	625	633	769	811	333	260
520	068	150	904	022	231	048	514	174	137	328	517	856
729	079	643	755	126	733	288	019	789	780	468	638	491
292	015	134	064	278	864	033	560	227	642	775	707	531

Chapter 7

The *z* Test

The *z* test is a gentle way to introduce you to important concepts of inferential statistics such as the sampling distribution, null hypothesis testing, and alpha. Its gentleness comes from not having to infer the population mean and population standard deviation; for the *z* test, they are both known. Where might you look to find population means and standard deviations? Several good sources are the Census Bureau, norms for standardized tests, and professional association databases.

Inferring population parameters adds another dimension of complexity to statistical procedures. Such complexity you will encounter in Chapter 8.

The research question being addressed by the *z* test is whether the sample is representative of the population. Representativeness is defined as the difference between the sample mean and the population mean, which computationally is expressed as the difference $M - \mu$. *This difference is called sampling error.* The smaller this difference, the closer the sample mean is to the population mean and the more representative the sample is of the population. The larger the difference, the farther apart the sample mean is from the population mean, and the less confident we are that the sample is representative of the population.

Is there a point at which the sample mean becomes so far removed from the population mean that you conclude the sample is not representative of the population? Yes, and determining where that point is will be covered in the section on alpha.

The Sampling Distribution of the Mean

At this point in the course, and I daresay at this point in your life, you have been exposed to only two distributions, the sample distribution containing raw scores and the population distribution of raw scores. Most students do not realize that there are a number of other distributions used in statistics. Several of these distributions you are encountering in your statistics course.

The *z* test requires using a distribution whose "raw scores" are sample means. You may need to think about such a distribution which is composed of sample means. This distribution is called the sampling distribution of the mean, and here is the story behind its creation.

Once upon a time, there was a population. From this population was randomly selected a sample of a particular *N*. The members of the sample were measured on the variable of interest (e.g., intelligence, resting pulse rate, number of words recalled), and the mean of the sample was computed and saved. The members of the sample were returned to the population, and another sample of the same *N* was randomly selected from the population. The sample mean was computed and saved, the members of the sample returned to the population, and yet another sample was selected and a sample mean computed. This procedure was repeated over and over until there were many sample means, let's say 500 (note: the theory assumes an infinite number of samples, but using a finite number for explanation is acceptable).

The 500 sample means are then converted into a frequency distribution, using the same procedure you learned in Chapter 3 to eyeball the data. What shape do you think the distribution of sample means takes? The shape is the normal distribution.

Critical Thinking Question: What do you expect the similarity to be between the sample mean and the population mean if the sample is randomly selected from the population? Remember that each sample is randomly selected from the population and random selection is the best way to ensure that a sample is representative of the population. Thus, you expect the sample mean and the population mean to be very similar to one another.

Critical Thinking Question: If a representative sample produces a sample mean which tends to be similar to the population mean, then what are you able to say about the size of the sampling error, $M - \mu$? That the size of the difference is small.

What do you expect when you compare the size of the 500 sample means with the size of the population mean? You expect to see most of the sample means very similar to the population mean, with a few sample means being more different than the population mean.

Critical Thinking Question: If you take the 500 means and add them together and divide the sum by 500, what do you expect the "mean of the sample means" to be? The population mean or μ.

Yes, the "mean of the sample means" is the population mean. This is quite important because it allows the sampling distribution of the mean to be a valuable tool for determining the sample's representativeness of the population.

Standard Error of the Mean. Does the sampling distribution have a standard deviation? Yes, it is called the standard error of the mean, symbolized as $\sigma_M \left(or\ \sigma_{\bar{X}} \right)$. The formula for the standard error of the mean is $\sigma_M = \dfrac{\sigma}{\sqrt{N}}$, where N is the size of the sample.

Examine this formula. Now look at the following fractions: $\dfrac{1}{2}, \dfrac{1}{3}, \dfrac{1}{4}, \dfrac{1}{5},$ and $\dfrac{1}{6}$. What is happening to the size of the overall value of the fraction as the denominator increases? As you can see from the sequence, the overall value of the fraction *is decreasing* as the denominator *is increasing*. The denominator in the formula for the standard error of the mean is the sample size, so when it increases, the standard error of the mean becomes smaller.

Critical Thinking Question: Is the sample more or less representative of the population as the sample N becomes larger? More representative.

Critical Thinking Question: Will the sample mean become more similar to or more different from the population mean as the sample N becomes larger? More similar.

Critical Thinking Question: As the sample N gets larger, does the standard error of the mean becomes larger or smaller? Smaller.

Critical Thinking Question: As the standard error of the mean gets smaller, does the sampling distribution become more leptokurtic or platykurtic? Leptokurtic.

As sample size increases, the random samples become more similar to the population and thus their sample means become more similar to the population mean. The standard error as the measure of variability decreases because of the greater similarity among the sample means themselves and between the sample means and the population mean. The resulting sampling distribution of the mean becomes more leptokurtic as sample size increases, and this makes perfectly good sense. Does it to you?

Before presenting the *z* test, there are two additional topics to cover first: null hypothesis testing and alpha.

Null Hypothesis Testing

Answering many of the questions in statistics requires using a procedure called null hypothesis testing. A synonym for hypothesis is prediction; the hypothesis states an expectation for the results of a study. The null hypothesis, symbolized as H_0, predicts no difference (hence the use of the word "null"), but no difference between what? Between the mean of the population *from which the sample was selected* and the mean of the population which you actually know. The null hypothesis predicts that the two population means are equal.

Critical Thinking Question: What does the null hypothesis predict about the sample being representative of the population? The sample is representative of the population.

Symbolically, the null hypothesis is written $H_0: \mu_1 = \mu$. μ_1 symbolizes the mean of the population from which the sample was selected (hence the subscript 1). We do not know this actual population mean of the population from which the sample was selected, so we use the sample mean to infer it. μ symbolizes the mean of the population, which we do know.

In contrast to the prediction of the null hypothesis is the prediction of the alternative hypothesis. The alternative hypothesis is symbolized as either H_1 or H_a. The alternative hypothesis predicts that the mean of the population from which the sample comes will be different from the actual mean of the population. This prediction is symbolized as $H_1: \mu_1 \neq \mu$.

After completing the statistical procedure and analyzing the results, the researcher must make one of two decisions:
• reject the null hypothesis OR
• fail to reject the null hypothesis (your textbook may call this decision *accepting the null hypothesis*).
 Once the decision is made, then the question of whether the sample is representative of the population can be answered.

Alpha

On what basis does the researcher decide to reject or not reject the null hypothesis? What "guide" is available for making this determination? The "guide" is alpha, and it serves several purposes.

Purpose 1. Alpha is the most extreme portion of the sampling distribution of the mean. An alpha of .05 consists of the most extreme 5% of the total area of the sampling distribution of the mean. This extreme 5% of the area is split equally between the two tails of the sampling distribution. Thus, the boundaries into this alpha area are ±1.96, which include the most extreme 2.50% of the area above the mean and the most extreme 2.50% of the area below the mean (2.50% + 2.50% = 5%).

Purpose 2. Alpha is the probability indicating the sample mean is no longer considered to be a part of the sampling distribution of the mean and thus no longer part of the population from which the samples in the sampling distribution of the mean have been randomly selected. If the sample mean is located in the alpha area of the sampling distribution, then the probability of it being a part of the sampling distribution of the mean is so low that you are more accurate to conclude that the sample comes from some other population with its own sampling distribution.

Purpose 3. Alpha is the probability of making a Type I error, which is incorrectly rejecting the null hypothesis when the null hypothesis is in fact true (more of Type I error in Chapter 9).

How does the researcher locate the sample mean on the sampling distribution to know whether it is in the alpha area or not? This question is answered in the next section.

Locating the Sample Mean on the Sampling Distribution of the Mean

The *z* score formula presented in Chapter 4, $z = \dfrac{X - M}{SD}$, is used to locate the raw score on the sample distribution of raw scores. Although usually not presented in a textbook, the *z* score formula used to locate the raw score on the population distribution of raw scores is $z = \dfrac{X - \mu}{\sigma}$.

Compare these two formulas. Although they are different from each other, they are *generically* the same formula: $z = \dfrac{\text{the raw score} - \text{the mean}}{\text{standard deviation}}$. This is the approach one needs to take to locate a raw score relative to all of the other raw scores in a distribution.

Critical Thinking Question: For the sampling distribution of the mean, what is the "raw score," the distribution's mean, and the standard deviation? As the sampling distribution of the mean is a distribution of sample means, the sample means are the "raw scores" of the distribution. The mean of the sampling distribution of the mean is the population mean μ. The standard deviation is the standard error of the mean σ_M. Substituting these symbols produces the *z* score formula for the sampling distribution of the mean:

$$z = \frac{M - \mu}{\sigma_M}$$

This formula converts the sample mean to its *z* score equivalent (expressed in standard errors of the mean) so that you can locate the sample mean on the sampling distribution of the mean. If the sample mean is in the alpha area, then you reject the null hypothesis and conclude that the sample is not representative of the population. If the sample mean is not in the alpha area, then you do not reject the null hypothesis and instead conclude that the sample is representative of the population.

NOTE: The previous paragraph conveys the essence of null hypothesis testing. You will see this approach used in all of the remaining statistical procedures covered in future chapters.

Example

The norms for the Grade Orientation Scale, used to assess student motivation, are a population mean of 73 and population standard deviation of 11. A sample of 49 introductory statistics students takes the scale and produces a mean of 69 and standard deviation of 4. Determine whether this sample is representative of the population. Use an α of .05.

- The *z* test formula is $z = \dfrac{M - \mu}{\sigma_M}$.

- *M* and μ are presented in the problem so the formula is now $z = \dfrac{69 - 73}{\sigma_M}$.

- The standard error of the mean is $\sigma_M = \dfrac{11}{\sqrt{49}} = \dfrac{11}{7} = 1.57$.

- Plugging in 1.57 produces a $z = \dfrac{69 - 73}{1.57} = -2.55$.

The sample mean is located 2.55 standard errors of the mean *to the left* of the mean of the sampling distribution of the mean.

You may need to read the last sentence again to fully understand it. With α of .05, the alpha area begins at ±1.96 standard errors of the mean. *Critical Thinking Question*: Is –2.55 located inside the alpha area? Yes. Thus, the null hypothesis is rejected. The sample is too far away from the population mean to be considered representative of the population.

Problems

1. Rejecting the null hypothesis for the *z* test means what about the representativeness of the sample to the population?

2. As sample size increases, what happens to the kurtosis of the sampling distribution of the mean and why?

3. If the sample mean is located in the alpha area, what decision is made about the null hypothesis?

For Problems 4 and 5: The population of intelligence test scores has a population mean of 100 and a population standard deviation of 15.

4. What is the standard error of the mean for the sampling distribution of the mean constructed of samples with $N = 48$?

5. Is the sampling distribution more leptokurtic or platykurtic if the sample size decreases to an $N = 36$?

6. The population mean is to the population distribution as _____ is to the sampling distribution of the mean.

    a. the sample mean   b. the standard error of the mean   c. μ d. the sampling error

7. What proportion of scores in a normal distribution are in the alpha area when $\alpha = .05$?

8. Deviation score is to _____ distribution as _____ is to sampling distribution of the mean.

    a. sample, sum of squares
    b. frequency, sample
    c. population, standard error of the mean
    d. sample, sampling error
    e. none of the above

9. _____ is to the standard error of the mean as sample mean is to standard deviation.

10. *SD* is to the sample distribution as _____ is to the population distribution.

11. Using the *z* Distribution Table at http://www.statsoft.com/textbook/distribution-tables/#z, what are the boundaries into the alpha area when $\alpha = .01$?

12. Researcher Problem: A group of researchers compare the results of a new edition of an intelligence test with its previous version using a sample of elementary students who attend a school located on a university campus. Most of the students are children of university students and university faculty.

They write their results and submit the manuscript for publication, but the editor returns the manuscript with a question: How do the authors know whether or not the sample of students is representative of the general population of elementary children when the students are children of university students and faculty? If the students in the study's sample are not representative, then the generalizability of the comparisons between the old and new intelligence tests is not possible, and the manuscript would not be accepted for publication. The editor indicated that without evidence to show that the sample of students is representative of the population of elementary school children, then the manuscript would not be published.

To answer the editor's question requires performing the *z* test. The researchers provide you with the following information: The population mean for the intelligence test is 100 and the population standard deviation was 15. For the sample of elementary students, the sample size was 25, and the mean of the sample was 105.

Was the manuscript published or not?

Chapter 8

The One-Sample *t* Test

The one-sample *t* test is used in two different situations. In one situation, the researcher has a sample and wants to determine whether it is representative of a population where the population mean is known (or hypothesized), but the population standard deviation is not. For example, some standardized tests report norms for the population mean but do not include the population standard deviation.

In the second situation, the researcher wants to determine the validity of the accuracy of an estimate for the population mean based on the sample. In this situation, both the population mean and the population standard deviation are estimated.

The one-sample *t* test parallels Chapter 7's *z* test in many ways. Those corresponding connections are presented in the table below. You will not understand some of the material as it has yet to be presented, but the table provides a good overview. The biggest difference between the two tests is that the population parameters are known in the *z* test but estimated in the *t* test.

z test	*t* test
M	M
μ	μ or μ_0 (the estimated population mean)
σ	s (the estimated population standard deviation; also the standard deviation of the sample used to estimate the population standard deviation)
σ_M	SE_M (the estimated standard error of the mean)
alpha	alpha
sampling distribution of the mean	sampling distribution of the mean
z	t
—	*df* (degrees of freedom)

Estimating the Population Standard Deviation

Perspective 1: Removing bias. For descriptive purposes, the standard deviation of a sample is

$$SD = \sqrt{\frac{SS}{N}} = \sqrt{\frac{\Sigma(X - M)^2}{N}}$$ (see, for example, Sprinthall, 2012). For estimation purposes, however, a change to this formula is necessary when you need to use the sample standard deviation to estimate the population standard deviation. Why is a change needed?

Critical Thinking Question: Do you recognize the expression $\Sigma(X - M)^2$ in the numerator of the *SD* formula? It is the sum of squares. Do you remember that the sample sum of squares is *smallest* when the sample mean is used? If you substitute any other number for the sample mean into the $\Sigma(X - M)^2$ expression, the resulting answer will be *larger than* the sum of squares.

Even with random sampling, there is a far greater likelihood that the sample mean is somewhat different from the population mean than identical to it. Since the population mean will tend to be different from the sample mean, the sample sum of squares computed using the sample mean is too small to be used as an estimate for population variability. Using the sample sum of squares will *underestimate* the population standard deviation. Thus, *SD* will likewise underestimate σ. An adjustment is needed to counteract the numerator being too small, but the numerator cannot be changed.

Critical Thinking Question: Looking at the *SD* formula and knowing that $\Sigma(X - M)^2$ cannot be changed, how can the formula be adjusted so that its overall value is larger? If it is not possible to adjust the numerator, the only option is to make the denominator smaller.

Making the denominator smaller increases the overall value of the formula's answer and thus counteracts the underestimation by the sample sum of squares. How small should the numerator become? Just a little bit smaller is necessary, specifically, $N-1$. The new formula is now

$$s = \sqrt{\frac{\Sigma(X - M)^2}{N - 1}}$$ where *s* is the symbol for the estimated standard deviation of the population.

Perspective 2: The Degrees of Freedom. The degrees of freedom are the number of scores in the sample which are "free" to take on any value when the mean of the sample is known. You have a sample of five young children's ages, and the mean age is 6. Jill's age is 3, Freddie's age is 8, Samantha's age is 5, and Jose's age is 5. If you add up these four ages, the sum is 21. Since you know the mean and the *N*, you also know that the sum of the scores in the data set is 30; $\Sigma X = M \times N$, so $30 = 6 \times 5$.

Critical Thinking Question: If ΣX is 30, and the sum of the four ages is 21, then what must the fifth age be? 9 years old. By knowing the mean and the *N*, one of the scores in the distribution is not free to take on any value. Thus, $N - 1$ is the formula for determining the degrees of freedom.

When estimating the population standard deviation based on the sample, only $N - 1$ members of the sample are free to take on any value and thus be reflective of the population. The score of the remaining member of the sample is locked because the mean of the sample is known.

The Estimated Standard Error of the Mean

For the *z* test, the standard error of the mean is computed as $\sigma_M = \dfrac{\sigma}{\sqrt{N}}$. Because the population standard deviation is now estimated, this formula needs to be changed to produce an estimate of the standard error of the mean. *s* is substituted for σ, and the resulting formula becomes $SE_M = \dfrac{s}{\sqrt{N}}$. SE_M is the symbol for the estimated standard error of the mean (your textbook may have it as $s_{\overline{X}}$).

The Null and Alternative Hypotheses

When the research question asks whether the sample is representative of the population whose mean is known, but standard deviation is not, then H$_0$: $\mu_1 = \mu$ where μ_1 is the mean of the population from which the sample was selected, and μ is the mean of the population which is known. The alternative hypothesis is then H$_1$: $\mu_1 \neq \mu$.

When the question is determining whether an estimated mean of the population could be viable given the sample, then H$_0$: $\mu_1 = \mu_0$, where μ_1 is the mean of the population from which the sample was selected and μ_0 is the estimated mean of the population. The alternative hypothesis is H$_1$: $\mu_1 \neq \mu_0$.

The *t* Distribution

For the one-sample *t* test, either one or two population parameters (that is, either the population standard deviation or the population standard deviation and the population mean) are being estimated based on the sample. *Critical Thinking Question*: What one key characteristic of the random sample influences how accurate the parameter estimates are? Sample size.

What influence does the size of the random sample have on how accurately it estimates the population? The larger the random sample (i.e., the larger the *N*), the more similar the sample is to the population and thus the more accurate are the estimates of the population parameters.

Critical Thinking Question: As a member of student government, you want to survey the student body to ascertain the critical issues. Will you have more confidence generalizing to the student body if 10 or 500 students respond to the survey? 500 students.

The accuracy of the estimates is dependent upon sample size, so sample size has to be taken into account when testing the null hypothesis for the one-sample *t* test. Sample size was not critical in the *z* test because no estimation of the population based on the sample occurred. You actually knew the population mean and population standard deviation when doing the *z* test.

How is sample size handled in the one-sample *t* test? William Gossett writing under the pseudonym Student developed a family of distributions he called the *t* distribution. The *t* distributions replace the normal distribution as the sampling distribution of the mean.

Like the normal distribution, the *t* distributions are unimodal, bell shaped, and symmetric around the mean. However, the shape of the distributions is also dependent upon the degrees of freedom, which is $N - 1$. For each degree of freedom, there is a differently shaped distribution. The *t* sampling distributions corresponding to smaller (i.e., less than 30) degrees of freedom have a rounder shape with more of the area in the tails as degrees of freedom gets smaller.

When the sample size is ∞ (this symbol means infinite), the *t* distribution is the same as the normal distribution used for the *z* test. This makes good sense because when the sample size is infinite, the sample IS the population. Thus, the sample mean and standard deviation are *perfect* predictors of the population parameters, and the normal distribution can be used.

As the sample size decreases, and thus as the degrees of freedom get smaller, the shape of the corresponding *t* sampling distribution becomes rounder and more platykurtic. Because more of the sampling distribution's area is pushed into the tails as *N* decreases, this pushes the alpha area's boundaries, called critical values, farther away from the sampling distribution's mean. For example,

• When alpha is .05 and the *z* sampling distribution OR the *t* distribution with infinite degrees of freedom is used, the alpha area begins at ±1.96; thus ±1.96 are the critical values

• When alpha is .05 and the *t* distribution for 10 degrees of freedom is used, the alpha area begins at ±2.228, thus ±2.228 are the critical values

• When alpha is .05 and the *t* distribution for 4 degrees of freedom is used, the alpha area begins at ±2.776, thus ±2.776 are the critical values

The following graph of these three sampling distributions shows the different shapes.

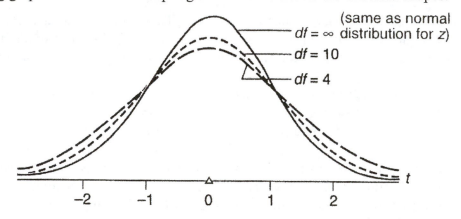

The boundaries into the alpha area are called critical values. They are being pushed farther away from the mean of the sampling distribution of the mean as degrees of freedom get smaller and the corresponding *t* distributions become flatter. This safeguard of the differently shaped *t* distributions makes rejecting the null hypothesis more difficult when sample sizes are small. Small samples have to produce larger differences before the null hypothesis is rejected.

The One-Sample *t* Test

Like the *z* test, determining whether the null hypothesis should be rejected or not requires the researcher to locate the sample mean on a distribution of sample means to see if that location is in the alpha area or not. Here are the previous formulas for locating scores on distributions; see if you can name the distributions:

$$\bullet \, z = \frac{X - M}{SD} \qquad\qquad \text{sample distribution of the raw scores}$$

$$\bullet \, z = \frac{X - \mu}{\sigma} \qquad\qquad \text{population distribution of the raw scores}$$

$$\bullet \, z = \frac{M - \mu}{\sigma_M} \qquad\qquad \text{sampling distribution of the mean}$$

For the one-sample *t* test, the formula is $t = \dfrac{M - \mu}{SE_M}$. This formula produces the calculated value of *t*, which is the location of the sample mean on the *t* sampling distribution of the mean.

Is this location in the alpha area or not? For finding that out, you need to consult the table of *t* values at http://easycalculation.com/statistics/t-distribution-critical-value-table.php OR the appendix of a textbook.

Find the *t* value which corresponds to the selected alpha level (usually .05), the number of tails (one or two, the difference will be explained in the next chapter), and the number of degrees of freedom. The negative and positive critical values are the boundaries into the alpha area. If the *t* you calculate from the formula exceeds the critical value *t* you find in the table, then you reject the null hypothesis. Either the sample is not representative of the population or the estimated mean of the population is not feasible given the sample.

Example

A sample of five seniors completes a questionnaire on attitudes about the death penalty. Scores on the questionnaire are 31, 32, 35, 36, and 39. Test the null hypothesis that this sample is representative of the population whose mean attitude toward the death penalty is 39 and use the two-tailed values. Use an alpha of .05.

• Adding all five scores produces a ΣX of 173.

• Squaring the five scores and then adding them produces a ΣX^2 of 6027.

• With an *N* of 5 and ΣX of 173, the *M* is 34.60.

• The sum of squares $= 6027 - \dfrac{173^2}{5} = 6027 - \dfrac{29{,}929}{5} = 6027 - 5985.80 = 41.20.$

• The estimated population standard deviation $= s = \sqrt{\dfrac{41.20}{4}} = 3.21.$

• The estimated standard error of the mean $= SE_M = \dfrac{3.21}{\sqrt{5}} = 1.44$.

• The calculated value of *t* is $\dfrac{34.60 - 39}{1.44} = -\dfrac{4.40}{1.44} = -3.06$.

• With 4 degrees of freedom and an alpha of .05, the boundaries into the alpha area are ±2.78.

• The sample mean of 34.60 is located 3.06 estimated standard errors of the mean to the left of the mean. This location is beyond –2.78 and thus in the alpha area.

• You reject the null hypothesis and conclude that the sample is not representative of the population mean attitude about the death penalty. Thus, the researchers cannot generalize from this sample to the population.

Problems

1. True or false: As sample size increases, the boundaries of the alpha area of the *t* sampling distribution of the mean move closer to the sampling distribution's mean.

2. Degrees of freedom is to sample size as 26 is to _____.

3. For the *t* distribution with 30 degrees of freedom and alpha set at .05, the boundaries of the alpha are _____.

4. For a sample of 18 subjects, the *t* values needed to reject the null hypothesis at the .01 level are ±2.898 what?

5. Increasing alpha moves the critical values of *z* or *t* _____ (closer to, further away) from the sampling distribution's mean.

6. Rejecting the null hypothesis becomes _____ (less, more) difficult as alpha decreases.

7. True or false: Changing alpha from .01 to .05 makes the alpha area smaller.

   Use the following sample of 92, 98, 100, 111, and 122 to answer problems 8-10.

8. Compute the estimated standard deviation of the population.

9. Compute the estimated standard error of the mean.

10. Using an alpha of .05, determine whether the sample is representative of a population whose mean is 108?

11. The null hypothesis is to the alternative hypothesis as no difference is to _____.

12. The critical values of *t* with infinite degrees of freedom and an alpha of .05 is to _____ as the critical values of *z* for alpha of .01 is to ±2.58.

13. Researcher Problem: A sample of 10 adults participated in a research study conducted by a pharmaceutical company testing a new antibiotic. For this study, the researchers have to make sure that the sample's average body temperature does not significantly differ from normal (i.e., 98.6°) so that the sample is representative of the population in order to generalize the sample's results to the population.   The researchers took the temperatures of each adult and they are reported below:

99.0, 97.0°, 98.6°, 97.7°, 98.3°, 98.0°, 98.6°, 98.8°, 97.9°, 98.1°

They have come to you in order to determine whether their sample is representative of the population or not. If the sample is representative, they can proceed with the research study. Can the researchers begin their study or not?

1. What are the H_0 and H_1?

2. What is *M*?

3. What is μ (it is in the problem but not obviously identified–do you see it)?

4. What is Σx^2?

5. What is *s*?

6. What is σ?

7. What is SE_M?

8. What is *t*?

9. What is α?

10. How many degrees of freedom?

11. What are the critical values?

12. Do you reject or fail to reject H_0?

13. Do the researchers begin their study or not?

The data for this Research Problem are given so that you can enter the data into statistical software such as SPSS or SAS to analyze them using the one-sample *t* test. Below are two different tables from the SPSS printout for this problem for you to use to assist you in answering the problems.

One-Sample Statistics

	N	Mean	Std. Deviation	Std. Error Mean
TEMPRTRE	10	98.2000	.59255	.18738

One-Sample Test

	Test Value = 98.6			
	t	df	Sig. (2-tailed)	Mean Difference
TEMPRTRE	-2.135	9	.062	-.40000

Chapter 9

The Two-Sample *t* Test for Independent Samples

The term *independent* in the two-sample *t* test for independent samples means that no score belonging to one of the samples is also a member of the other sample. No score is a member of both samples.

The *z* test and the one-sample *t* test compare a sample mean and a population mean. In contrast, the two-sample *t* test compares the means of two samples. The null hypothesis for this statistical test is $H_0: \mu_1 = \mu_2$. What is this null hypothesis expressed in words? The mean of the population from which the first sample was selected equals the mean of the population from which the second sample is selected.

The alternative hypothesis is $H_1: \mu_1 \neq \mu_2$, which says that the mean of the population from which the first sample is selected is different from the mean of the population from which the second sample is selected.

How do you test the null hypothesis? The critical tool you need is a sampling distribution, and the one for testing this null hypothesis is different from the sampling distribution of the mean.

The Sampling Distribution of Differences

Time for a story that is a bit more complex than the one in Chapter 7. Once upon a time there was a population. From this population were randomly and simultaneously selected two samples, both with the same *N,* such that no member of once sample was a member of the other sample. The members of each sample were measured, and the means for each sample were computed.

Critical Thinking Question: The scores were randomly selected from the same population. Do you expect the two sample means to be more similar or more different from one another? More similar even though no member of one sample is a member of the other sample.

You then take the two sample means and subtract one from the other. Do you expect the difference to be close to 0 or different from 0? The difference will tend to be close to 0.

Once you have the difference between the two sample means, save it, and return the members of both samples to the population. Then randomly and simultaneously select another two samples, both with the same *N* as the first pair. Measure the members of both samples, compute the means, and subtract one sample's mean from the other. Save the difference and return the members to the population. Repeat the process over and over until you have, let's say, 500 differences between sample means (note: the theory assumes an infinite number of differences between sample means, but using a finite number for explanation is acceptable).

Imagine yourself examining these 500 differences. What do they look like? What do you expect the most common difference to be? If you graph the differences as a frequency distribution, what pattern do they form?

The 500 differences form a normal distribution. The mean of the distribution is 0, which is also the mode or most frequent difference in the data set (remember that the mean, median, and mode are the same value in the normal distribution). Most of the differences are close to 0, one–half of them negative values and one–half of them positive values. A few of the positive and negative differences are larger. The larger the differences are, the *less* frequently they occur.

This distribution is called the sampling distribution of differences. You know by now that the mean of any distribution expressed in standard deviation units (i.e., as a z score or t score) is always 0. For the sampling distribution of differences, the mean expressed in the *actual* measurement units (e.g., pound, inches, points on a test) is also 0. The standard deviation of the sampling distribution of differences is called the standard error of difference. Because the population standard deviations of the two samples are estimated, you will also compute the estimated standard error of difference.

Are you ready for this: The estimated standard error of difference is the square root of the sum of the two estimated variances of the sampling distributions of the mean for each of the two samples. The symbol version of the formula might be easier for you to understand than the text version (would this have been true for you at the beginning of the course?):

$$\text{estimated standard error of difference} = SE_D = \sqrt{SE_{M_1}^2 + SE_{M_2}^2}$$

Remember from Chapter 4 that the standard deviation^2 is the variance. $SE_{M_1}^2$ is computed by squaring the estimated standard error of the mean of the first sample.

The above formula computes the separate variance estimate of the standard error of difference. When the Ns of the two samples differ, use the pooled variance estimate formula:

$$SE_D = \sqrt{\frac{(N_1 - 1)s_1^2 + (N_2 - 1)s_2^2}{N_1 + N_2 - 2}}$$

The t Formula

The t formula for locating the difference between the two sample means on the sampling distribution of differences follows the generic pattern $\dfrac{\text{raw score} - \text{the mean}}{\text{standard deviation}}$.

• What are the "raw scores" of the sampling distribution of differences?
   The differences between the two sample means

• What is the mean of the sampling distribution of differences? 0

• What is the symbol for the estimated standard error of difference? SE_D

So the t formula for the two-sample t test is:

$$t = \frac{(M_1 - M_2) - 0}{SE_D}, \text{ which is shortened to } t = \frac{M_1 - M_2}{SE_D}$$

Your textbook might present the formula as $t = \frac{M_E - M_C}{SE_D}$, where E stands for the experimental group and C stands for the control group.

Degrees of Freedom

Degrees of freedom for the two-sample t test are computed by adding $(N_1 - 1)$ for the first sample and $(N_2 - 1)$ for the second sample. This formula can also be expressed as $N_1 + N_2 - 2$.

The Scientific Method and Experimental Control

The most basic experimental research design involves two groups of participants. One of the groups is exposed to some treatment and the other group is not. The group exposed to the treatment is called the experimental group; the group not exposed to the treatment is called the control group.

Two important terms introduced on p. 39 at the beginning of Chapter 5 reappear here: *independent variable* and *dependent variable*. The independent variable is called independent because it can be freely manipulated by the experimenter to form at least two groups or treatment conditions. For the two-sample t test, the independent variable produces two groups, usually resulting from the *presence or absence of the treatment*. Participants in the experimental group receive the treatment, whereas participants in the control group do not (i.e., they are given a placebo).

The dependent variable is the measure the researcher employs to "detect" the effect of the independent variable. It is called the dependent variable because its values depend on the independent variable. The phrase $Y = f(X)$ also appeared on p. 39 in Chapter 5. It means that the values graphed on the Y axis are a function of, or dependent upon, the values graphed on the X axis. Hence, the dependent variable is graphed on the Y axis, and the independent variable is graphed on the X axis. **HELPFUL HINT**: Applying this information to a graph in a research article can help you verify what the independent and dependent variables are.

The treatment can be anything the researcher wants to test in order to determine its effectiveness or role. A new drug, a different way of teaching a course, gender, personality, a new strategy for motivating workers, or a new fertilizer for wheat are a few examples.

IMPORTANT: Concluding that a treatment is effective requires equating or keeping constant all the characteristics of the two groups which could affect the measurement of the experiment's dependent variable *except* that one group receives the treatment and the other group does not. Usually, this constancy between groups is achieved through randomly assigning participants to either the experimental or control group.

To the extent that the constancy is achieved, then the experiment has high *internal validity*.

Internal validity means that any statistically significant difference between the means of the experimental and control groups is explained only by the treatment and nothing else. Thus, the treatment *causes* the difference in the dependent variable. If there are other variables on which the two groups differ, these *extraneous variables* become rival explanations for why the two groups are statistically significantly different. The researcher's goal is to have NO extraneous variables and thus high internal validity.

Assumptions to Use the *t* Test

The three assumptions for the *t* test (Sprinthall, 2012) are:
1. The scores in each group are normally distributed.
2. Population variances are equal for the two groups (also known as homogeneity of variance).
3. Each participant's score is not affected by other participants in the same group

Example

A researcher wanted to study the effect of a new medication for reducing fever. She enlisted five doctors to assist. 50 adult patients hospitalized by the five doctors were randomly given either the new medication or a placebo. The 25 patients receiving the new medication belonged to the experimental group. The 25 patients receiving the placebo belonged to the control group. Two tablets of the new medication or two tablets of the placebo were administered at four hour intervals. The dependent variable was the change in body temperature, defined as (the body temperature after 12 hours of medication – the temperature at admission).

All patients' temperatures were recorded at admission and every four hours thereafter. After the first 12 hours, the experimental group's mean temperature change was $-1.46°$ with an *s* of 1.33. The control group's mean temperature change was $-.45°$ with an *s* of .94. Is the difference between the experimental and control groups' mean changes significant?

To answer the question, you compare the two groups' means using the independent–samples *t* test. If $t = \dfrac{M_E - M_C}{SE_D}$, then $t = \dfrac{-1.46 - (-.45)}{SE_D} = \dfrac{-1.46 + .45}{SE_D} = \dfrac{-1.01}{SE_D}$.

What is SE_D? The *s* for each group is given. Each *s* must be divided by the square root of *N* to compute the estimated standard error of the mean SE_M:

$$SE_{M_E} = \frac{1.33}{\sqrt{25}} = .27, \text{ and } SE_{M_C} = \frac{.94}{\sqrt{25}} = .19,$$

To find the estimated standard error of difference, plug into the formula the SE_{M_E} and SE_{M_C}:

$$SE_D = \sqrt{SE_{M_E}^2 + SE_{M_C}^2} = \sqrt{.27^2 + .19^2} = .33$$

Thus, $t = \dfrac{-1.01}{.33} = -3.06$. The difference of $-1.01°$ between the two mean body temperature changes of $-1.46°$ and $-.45°$ is located 3.06 estimated standard errors of difference to the left of the mean of the sampling distribution of differences.

With an alpha $= .05$, is this location in the alpha area? To find out, you go to the table in the appendix of your textbook where the critical values are given. How many degrees of freedom are there? $(25 - 1) + (25-1) = 50 - 2 = 48$.

The critical values or boundaries into the .05 alpha area for the sampling distribution of differences corresponding to 48 degrees of freedom are located at ± 2.011 estimated standard errors of difference away from the mean. The difference between the two sample means is located at -3.06 estimated standard errors of difference. Is the location of the difference between the two samples in the alpha area or not? Yes, -3.06 is located in the alpha area.

What decision is then made about the null hypothesis? To reject the null hypothesis. The two means are statistically different from each other. What caused the difference? If the two groups differ *only* on the experimental group receiving the new medication, then the new medication must have caused the difference.

One–Tailed versus Two–Tailed t Test

Look at the formulas for the z test, the one-sample t test, and the two-sample t test. All have the standard deviation in the denominator, and the standard deviation can only be positive. Thus, the sign (i. e., $+$ or $-$) of the calculated z or t value is determined by the difference in the numerator of the formula.

Since the calculated z or t can be either negative or positive, you have to be prepared with having a part of alpha on both sides of the sampling distribution. Specifically, you need to split alpha so that one–half of the alpha area is located in the negative tail of the sampling distribution, and the other one–half of the alpha area is located in the positive tail of the sampling distribution. You literally need to "cover your tails" because the calculated value of z or t could be positive or negative. You needed to be ready for either outcome.

Different Alternative Hypotheses. The alternative hypothesis for the two-sample t test is H_1: $\mu_1 \neq \mu_2$. This alternative hypothesis predicts that the two groups differ from each other, and it does not matter whether the inequality is a result of one mean being larger than or smaller than the other. This alternative hypothesis is called a *nondirectional alternative hypothesis*.

Suppose, however, that based on previous evidence, the researcher is reasonably confident that the treatment will have an effect. The researcher can then specify that one group will have a mean score greater than the other group's. In the example about temperature change, the researcher predicts that the temperature loss is greater for the experimental group compared to the control group: H_1: $\mu_E < \mu_C$. This alternative hypothesis specifying the direction of the inequality is called a *directional alternative hypothesis*. It specifically defines the direction of the difference between the two sample means.

Critical Thinking Question: Do you understand how greater temperature loss by the experimental participants translates into an H₁ where M_E is less than M_C? Temperature loss is a negative number. The greater the temperature loss, the farther in the negative direction the loss is. Remember from the number line on p. 6 in Chapter 1, values to the left are always smaller than values to the right. If the new medication is effective, you would expect the experimental group's mean to less than the control group's mean.

Critical Thinking Question: Which is the more precise statement:
• the height of a 5' person is different from the height of a 6' person
• the height of a 5' person is shorter than the height of a 6' person
The second statement is more precise because the direction of the inequality is specified.

Examine the directional alternative hypothesis H₁: $\mu_E < \mu_C$. What does this alternative hypothesis predict the sign of the difference M_E - M_C to be? The difference would be negative.

If the difference M_E - M_C is negative, then what is the sign of the calculated value of *t*? It is also negative, which means that the location of the difference between M_E and M_C is predicted to be on the *left* side of the sampling distribution of difference. This predicted location means that the one–half of alpha on the right side of the sampling distribution is wasted.

By making a directional alternative hypothesis, the researcher can "recycle" the wasted one–half of alpha by moving it to the opposite side of the sampling distribution and combining it with the other half of alpha. Thus, both halves of alpha are consolidated in one tail of the sampling distribution. The resulting *t* test is called a one-tail *t* test.

For the example of temperature change, when alpha is .05 and there are 48 degrees of freedom, the boundary into the alpha area moves from ±2.011 to –1.68 estimated standard errors of difference. The *t* distribution critical values table for showing both the one-tailed and the two-tailed *t* test is at http://easycalculation.com/statistics/t-distribution-critical-value-table.php or the back of your textbook.

Critical Thinking Question: When alpha is consolidated in one tail instead of split in two tails, does the boundary of the alpha area move closer to the mean of the sampling distribution of differences or farther away? Closer to the sampling distribution mean.

When researchers can articulate the direction of the alternative hypothesis, they realize a benefit from being more specific. The directional alternative hypothesis allows the one–tailed *t* test to be used. This test makes rejecting the null hypothesis easier because the boundary to the alpha area is closer to the sampling distribution mean.

Moving the boundary to the alpha area closer to the sampling distribution's mean means that a smaller difference between M_E and M_C is statistically significant. The treatment does not have to produce as large a difference for the difference to be statistically significant. The one–tailed *t* test thus makes it easier to reject the null hypothesis. Why does the researcher want to take advantage of such an opportunity? The answer is in the next section.

The Tension between the Null and Alternative Hypotheses

The null hypothesis predicts that the new treatment is not effective. Think of the null hypothesis as protecting the *status quo*. What is the status quo? It is the way things are right now without the possibility of change that the new treatment represents. In contrast, the treatment represents something new; it is a "threat" to the status quo. If the treatment is effective, then it should be implemented, and the status quo consequently must be changed.

In inferential statistics, the null hypothesis is tested by the statistical procedure (i.e., the *z* test or the *t* test). The procedure is designed to make rejecting the null hypothesis difficult. In the legal system, the defendant is assumed innocent until proven guilty. In research, the treatment is assumed not effective until proven effective.

The researcher, on the other hand, proposes the alternative hypothesis (also called the scientific hypothesis), which predicts that the treatment is effective. Researchers constantly search for better ways to do things, and the alternative hypothesis predicts a better way.

The alternative hypothesis represents change, thereby threatening the status quo. The statistical procedure protects the status quo by making the rejection of the null hypothesis difficult. The pressure is on the researcher to present convincing evidence about the treatment's effectiveness to overcome the procedure's built-in bias toward the null hypothesis and the status quo.

Type I and Type II Errors

Whenever a statistical decision is made about rejecting or failing to reject the null hypothesis, even when great care has been taken in carefully designing an experiment and applying the correct statistical procedure, there is some probability that the decision is the wrong one. The goal is to minimize that error.

In statistics, the decision is whether to reject or not reject the null hypothesis. The Type I error occurs when the researcher decides to reject the null hypothesis when the null is in fact true. Thus, the treatment is erroneously found to be effective when it is actually not effective. The probability of making this error is alpha. Thus, if alpha is .05, then the probability of rejecting the null hypothesis when the null is true is .05.

The Type II error is not rejecting the null hypothesis when the null is false. The treatment is erroneously found to be ineffective when it actually is. Beta is the probability of making Type II error. The probabilities of making alpha and beta errors are inversely related—as the probability of one increases, the probability of the other decreases.

Critical Thinking Question: Which error is the more serious to commit when testing the null hypothesis? It depends on the null hypothesis. For example, if the null hypothesis predicts that the experimental medication is no different than the placebo, committing the Type I error means the null is true, but the researcher erroneously concluded from the results that the drug is effective when in fact it is not.

More broadly speaking, rejecting this null when the null is true means that the researchers are recommending that a new practice, a new treatment, or some other change should be adopted when in fact it should not. The likelihood of harm is greater when adopting an ineffective practice.

In contrast, the Type II error means that even though an opportunity for improving the status quo has been missed, the status quo just continues so you "keep on keepin' on"—no better but no worse. In null hypothesis decision making, when the null predicts that the treatment is not effective, the likelihood of harm is greater with the Type I than the Type II error. That is why introductory statistics textbooks emphasize Type I error and alpha in contrast to Type II error and beta.

For null hypotheses that predict that there is no problem or no concern, however, the Type II error can be more devastating. For example, the null hypothesis predicts that the patient does not have breast cancer. After analyzing the tissue from a biopsy, the medical laboratory concludes that a woman has breast cancer; however, in fact she does not. The null is rejected when in fact it should not be. This *false positive* is an example of a Type I error. Although this error creates tremendous anxiety before the proper diagnosis is made, consider the far greater life-threatening error of telling a woman she does not have breast cancer when in fact she does. This *false negative* mistake is Type II error and would result in the woman delaying important treatment. You can imagine the devastation to someone (and their significant other) who is HIV+ but the test results indicate the false negative.

Here is yet a different situation—Consider the fire department's decision when a report of a fire is called in. Not sending trucks to respond to this call would be unthinkable so the Type II error is never made. On the other hand, sending trucks when there is no fire would be a Type I error. *Critical Thinking Question*: What do we call this error? A false alarm.

In our legal system, the person is innocent until proven guilty. A jury can make one of two errors in its decision making: finding the innocent person guilty, which is the Type I error called false guilt, or finding the guilty person not guilty, which is the Type II error called false innocence. The first error is called false guilt and the second error is called false alarm. *Critical Thinking Question*: Which error is more serious? In the context of the American legal system, the Type I error of finding the innocent person guilty is the more serious error.

Get a load of this—In the conservatively cautious culture of statistics, it is more prudent and thus *better* to miss an opportunity to improve than it is to change and run the risk of that change causing harm. Although we do not tend to employ this view in our own lives, such an outlook serves to protect us. It helps to ensure that change (e.g., implementing a new treatment) occurs only when a history of supportive evidence clearly and conclusively indicates that the benefits of the change far outweigh the risks. It is for this reason that in science in general and especially in the medical/health and behavioral sciences, there are many replications of experiments whose results uniformly support change before the change is actually implemented as practice.

The next page presents information about Type I and Type II errors in a different way to help you understand the difference between these errors.

Table 9-1. Type I and Type II Errors

Null Hypothesis Decision Making

	In reality, the H_O is	
Researcher's Decision:	True	False
Do not reject H_O	Good Decision *Confidence Level* (1-alpha)	*Type II Error* (beta)
Reject the H_O	*Type I Error* (alpha)	Good Decision *Power* (1-beta)

Jury Decision Making

	In reality, the person is	
Jury's Decision:	Innocent	Guilty
Not guilty	Good Decision *Confidence Level* (1-alpha)	False Innocence *Type II Error* (beta)
Guilty	False Guilt *Type I Error* (alpha)	Good Decision *Power* (1-beta)

Fire Department Decision Making

	In reality, there is	
Fire Department's Decision:	No Fire	A Fire
Do not send fire trucks	Good Decision *Confidence Level* (1-alpha)	Miss *Type II Error* (beta)
Send fire trucks	False Alarm *Type I Error* (alpha)	Good Decision *Power* (1-beta)

Problems

Use the following scenario for Problems 1 through 4. The experimental group has a $M = 20$, $SS = 250$, and $N = 26$. The control group has a $M = 24$, $SS = 300$, and $N = 26$. Assume a directional alternative hypothesis using a one–tailed *t* test with the alpha = .05.

1. What is the estimated standard error of difference?

2. What is the calculated value of *t*?

3. What is the critical value?

4. Should the null be rejected or not?

   For Problems 5 and 6 indicate whether the hypothesis is null or alternative. If alternative, indicate whether it is directional or nondirectional.

5. The new technique for teaching statistics will improve student learning.

6. Men and women will not differ on their final grades in statistics.

7. Indicate which of the following increases the likelihood of rejecting the null hypothesis (there can be more than one answer):
   a. increasing alpha
   b. decreasing sample size
   c. using a one–tailed instead of two–tailed *t* test

8. What do we call the probability of incorrectly rejecting a true null hypothesis?

9. Is alpha error or beta error more serious and why?

10. When is a one–tailed *t* test warranted?

11. A two–tailed *t* test is to a _____ alternative hypothesis as a one–tailed *t* test is to a _____ alternative hypothesis.

12. Alpha is to Type _____ error as beta is to Type _____ error.

13. Sample mean is to the sample distribution as 0 is to _____.

14. The mean of sampling distribution of differences is to the standard error of difference as the mean of the population is to the _____.

15. Researcher Problem: The Campus Bookstore is selecting and ordering school spirit apparel for the new academic year in the colors of the university, black and gold. The bookstore manager is wondering whether students will prefer to buy clothing (e.g., t shirts, hoodies) that is colored more

with black or clothing that is colored more with gold.

The manager contacts you to conduct a research project to determine which color is more preferred by the student body. You identify a random sample of 34 students and ask them to respond to two items:
1. Which color do you prefer: black or gold?
2. On a scale of 1 (very low school spirit) to 5 (very high school spirit), rate your school spirit.

Using the logic that students with greater school spirit will buy more clothing that supports the institution, the data are used to answer the research question: Does the amount of school spirit differ by color preference? Based on the answer to this question, the manager will have information with which to base the decision to buy more of one color clothing than the other color.

The descriptive data are as follows in this computer printout:

Group Statistics

	COLOR	N	Mean	Std. Deviation	Std. Error Mean
School Spirit	Black	17	3.24	.903	.219
	Gold	17	2.47	.943	.229

Assuming a nondirectional alternative hypothesis, an alpha of .05, and a two-tail *t* test, determine what the manager should decide to do in selecting colors for next year's apparel order.

1. What are the H_0 and H_1?

2. What are M_1 and M_2?

3. What are s_1 and s_2?

4. What are SE_{M_1} and SE_{M_2}?

5. What is s?

6. What is σ?

7. What is SE_D?

8. What is t?

9. What is α?

10. How many degrees of freedom?

11. What are the critical values of t?

12. Do you reject or fail to reject H_0?

13. Should the manager order more clothing in one color rather than the other?

Here are the raw data if you want to analyze the Researcher Problem data with SPSS or some other statistical software:

Black	3
Black	3
Black	3
Black	3
Black	4
Black	3
Black	4
Black	3
Black	4
Black	3
Black	4
Black	5
Black	3
Black	2
Black	3
Black	1
Black	4
Gold	4
Gold	2
Gold	2
Gold	1
Gold	2
Gold	3
Gold	4
Gold	3
Gold	4
Gold	2
Gold	1
Gold	2
Gold	2
Gold	3
Gold	2
Gold	2
Gold	3

Chapter 10

The One–Way Analysis of Variance

If you look at the null hypotheses of the *z* test, one-sample *t* test, and two-sample *t* test, how many groups are being compared? Two groups are being compared. In contrast, the analysis of variance was developed to compare three or more groups. Why was a new technique needed?

The example in Chapter 9 for illustrating the two-sample *t* test compared two groups of patients: the experimental group received the new medication, and the control group received a placebo. Now the researcher wants to replicate (or repeat) the experiment but add a third group of 25 patients who receive the regularly prescribed aspirin the hospital staff normally administers for fever.

Using the two-sample *t* test approach for comparing two means, you compare the means of the three groups by pairs, resulting in three null hypotheses to test:

- $H_0: \mu_{\text{new drug}} = \mu_{\text{placebo}}$

- $H_0: \mu_{\text{aspirin}} = \mu_{\text{placebo}}$

- $H_0: \mu_{\text{newdrug}} = \mu_{\text{aspirin}}$

This approach seems reasonable, but it creates a problem. Repeated *t* tests on the same data set increase alpha to unacceptably high levels. What does this mean? It means that the increased size of alpha is also greater probability of committing a Type I error whereby rejecting the null hypothesis when the treatment is not effective becomes more likely. Thus, one of these null hypotheses could be rejected wrongly. The analysis of variance solves this problem.

Two Data Sets in One

With the addition of the third group, we now have one data set which is the sample of 75 patients. The sample is divided into three treatment conditions each with 25 patients: One treatment group receiving the new medication, one treatment group receiving the placebo, and the third treatment group receiving the regularly prescribed aspirin.

What is the second data set? The second data set is the sample of three treatment condition means. You may need to reread this sentence to fully appreciate it. For the analysis of variance, one data set is made up of the raw scores. A second data set contains the means of the treatment conditions.

The Null Hypothesis

The one–way analysis of variance, often referred to as the one–way ANOVA, simultaneously compares all the means using one null hypothesis. The null hypothesis for the example is $H_0: \mu_1 = \mu_2 = \mu_3$ because there are three treatment conditions. A fourth treatment condition, for example, μ_4 would result in an addition to the null hypothesis.

What's the Logic, Part I?

Critical Thinking Questions: What does the null hypothesis predict about the effectiveness of the three treatments? Does the null predict that the new medication will be better than the placebo? Does the null predict that the aspirin will be better than the placebo? The null hypothesis predicts that all three treatment condition means will have the same means. Neither the new medication or the aspirin will be effective because patients receiving these treatments will do no better than patients receiving the placebo.

Critical Thinking Questions: How would the means of the three treatment conditions compare with one another if the new medication is more effective than the aspirin and placebo and the aspirin is also more effective than the placebo? The three treatment condition means would be different from one another. Is it possible for two of the three treatment condition means to be statistically significantly different from one another but not from the third condition mean? Yes.

If one or both treatments are effective, then the means will be spread out from one another. If the treatments are not effective, then the means will be close to one another. Is it possible to measure this spread of the treatment condition means from one another? Yes What is part of Chapter 3's title? Measuring the spread.

How is spread measured? Let's review the steps from Chapter 3 for computing the sum of squares deviationally:

• The mean of the distribution is subtracted from each member of the distribution
• The differences are squared
• The squared differences are added together

For the data set containing the three treatment condition means, the mean has to be computed. This mean, symbolized as M_t, is called the total mean and is computed as $\dfrac{M_1 + M_2 + M_3}{3}$. This formula assumes that the three treatment means come from groups where the *n*s are equal. Such is the case in the example as each condition has 25 patients.

ASIDE: If the groups had unequal *n*s, then a weighted mean would need to be computed using the formula for computing the weighted means from the end of Chapter 3. The formula for combining means from treatment conditions with different *n*s is

$$M_t = \frac{(M_1 \cdot n_1) + (M_2 \cdot n_2) + (M_3 \cdot n_3)}{n_1 + n_2 + n_3}$$

To find the sum of squares of the data set of the three treatment condition means, you subtract the total mean from each of the treatment condition means, square each difference, and add them together:

$$\left(M_1 - M_t\right)^2 + \left(M_2 - M_t\right)^2 + \left(M_3 - M_t\right)^2$$

The data set containing the treatment condition means will eventually be compared to the second data set composed of raw scores. Because the treatment condition means and the raw scores are not similar measures, each of the three differences between means needs to be multiplied by the respective *n*s of the treatment conditions. These computations bring the means to the level of the raw scores. Thus, the complete expression becomes the formula for the *sum of squares between groups*:

$$\left(M_1 - M_t\right)^2 n_1 + \left(M_2 - M_t\right)^2 n_2 + \left(M_3 - M_t\right)^2 n_3$$

Critical Thinking Question: From Chapter 3, what do you get when you divided the sum of squares by the *N*? The variance.

Critical Thinking Question: Now that you are working in inferential statistics, what is the *N* replaced by? Degrees of freedom. What is the formula for degrees of freedom? $N - 1$. So for a data set containing three treatment means, how many degrees of freedom are there? 2.

Since each member of the first data set is a treatment condition mean rather than a raw score, there is a new symbol to represent the number of treatment means, and it is *k*. So the degrees of freedom for the data set containing the treatment means is $k - 1$.

The following expression is called the *variance between groups* or the *mean square between groups*:

$$V_b = MS_b = \frac{SS_b}{df} = \frac{\left(M_1 - M_t\right)^2 n_1 + \left(M_2 + M_t\right)^2 n_2 + \left(M_3 + M_t\right)^2 n_3}{k - 1}$$

What's the name of the statistical procedure presented in this chapter? The analysis of variance. The *variance between groups* is one of the two variances included in the analysis. The other variance is called the *variance within groups* and is presented in the next section.

What's the Logic, Part II?

Suppose that for the 25 patients receiving the new medication, 15 have their fevers reduced, 5 experience no change, and 5 have their fevers increased. While the overall mean indicates that the fever has gone down, quite a few of the patients either experienced no change or got worse.

This scenario introduces a new dimension to the meaning of the term *effective*. One definition of a treatment's effectiveness is that the mean of the group getting the treatment is different from the mean of the group not getting the treatment. The mean square between groups measures this.

Effectiveness also means that the treatment is uniformly effective—that participants exposed to the same treatment will respond in the same way to the same degree. When such consistent responsiveness to the treatment by those exposed to the treatment does not occur, then the treatment's effectiveness is reduced.

Critical Thinking Questions: What if all of the patients in the group receiving the new medication reduce their fevers by the same amount (i.e., all raw scores are the same):
• Does this result mean that the medication's effect is consistent or inconsistent across patients? Consistent
• How does the mean of the group compare with the reductions of the 25 clients?   The mean will be the same as the raw scores (all the raw scores are the same).
• Is the standard deviation of this group large or small? The standard deviation will be 0 as all scores are the same.

How can the consistency part of defining effectiveness be measured? Once again, the sum of squares is the answer. For the analysis of variance, the sums of squares for each of the three treatment conditions are computed and then added to one another. Using the deviational formula for the sum of squares from Chapter 3, the formula for the *sum of squares within groups* is:

$$\Sigma(X_1 - M_1)^2 + \Sigma(X_2 - M_2)^2 + \Sigma(X_3 - M_3)^2$$

This sum is then divided by the degrees of freedom to produce the *variance within groups* or the *mean square within groups*. The degrees of freedom for each of the treatment conditions is defined as $n - 1$. The overall degrees of freedom are $(n_1 - 1) + (n_2 - 1) + (n_3 - 1)$. This expression is shortened to $N - 3 = N - k$. The following expression is called the *variance within groups* or the *mean square within groups*:

$$V_w = MS_w = \frac{SS_w}{df} = \frac{\Sigma(X_1 - M_1)^2 + \Sigma(X_2 - M_2)^2 + \Sigma(X_3 - M_3)^2}{N - k}$$

The Analysis of Variance Formula

The analysis of variance is calculated as the ratio between the mean square between groups and the mean square within groups. This ratio is called the *F* ratio, the *F* honoring Sir Ronald Fisher, who invented the analysis of variance technique. Examine the formula below:

$$F = \frac{V_b}{V_w} = \frac{MS_b}{MS_w} = \frac{\frac{SS_b}{df}}{\frac{SS_w}{df}} = \frac{\frac{(M_1 - M_t)^2 n_1 + (M_2 - M_t)^2 n_2 + (M_3 - M_t)^2 n_3}{k - 1}}{\frac{\Sigma(X_1 - M_1)^2 + \Sigma(X_2 - M_2)^2 + \Sigma(X_3 - M_3)^2}{N - k}}$$

The *F* Sampling Distribution

It is story time again: Once upon a time there was a population. Two samples were randomly selected from this population. The sizes of the two samples could be the same OR they could have different *N*s. The members of both samples were measured on the variable of interest, and then the variances of the two samples was computed. Next, the variance of one sample was divided by the variance of the other sample. This ratio of the two variances was saved, the members of the two samples returned to the population, and the procedure repeated. The next pair of samples had the same *N*s as the first two samples. Their members were measured, the variances calculated, and the ratio between the two variances computed. Over and over the procedure was repeated for 500 times, each time producing a ratio of two variances.

Critical Thinking Question: Are any of these ratios negative? No, because variance can never be negative, the ratio of two variances can never be negative.

Critical Thinking Question: Two samples are randomly elected from the same population. In the last chapter, the difference between the means of these pairs of samples was usually what? 0 or very close to 0 because the two samples, although independent of each other, had similar means because they were randomly selected from the same population.

Critical Thinking Question: If the differences between two sample means randomly selected from the same population tends to be 0, what do you expect the ratio of their sample variances to be? 1. The two variances are similar estimates of the variance of the population from which they were selected. Thus, anything divided by itself is 1.

These 500 ratios are then converted into a sampling distribution called the F sampling distribution. Like the t distribution, the F distribution is a family of distributions. However, the F distribution has two different degrees of freedom—one for the variance in the numerator $(k-1)$ and one for the variance in the denominator $(N-k)$.

For an alpha of .05, the critical value of F is the point on the far right side of the abscissa which divides the total area of the distribution into 95% on the left side of the critical value and 5% on the right side. The critical values of F can be found in a table located at http://www.statsoft.com/textbook/distribution-tables/#f05 for $\alpha = .05$ and http://www.statsoft.com/textbook/distribution-tables/#f01 for $\alpha = .01$ or the back of your textbook.

Testing the Null Hypothesis

The F formula produces the calculated value of F. The table in the back of your statistics textbook or at http://www.statsoft.com/textbook/distribution-tables/#f05 produces the critical value of F for $\alpha = .05$. If the calculated value of F exceeds the critical value, then the null hypothesis is rejected. If the calculated value of F does not exceed the critical value, then the null hypothesis is not rejected.

Using the F Formula to Understand Methodology

If scientists predict that the treatment will be effective, then they are interested in seeing the null hypothesis rejected. Why? What does the null hypothesis predict about the effectiveness of the treatment? The null hypothesis predicts that the treatment will not be effective.

Rejecting the null hypothesis requires having a large calculated value of F. Look at the F formula—it is a fraction divided by a fraction. Thus, there is a numerator of the numerator, a denominator of the numerator, a numerator of the denominator, and a denominator of the denominator. What needs to happen to each of these four parts to increase the size of the F?

To increase the size of the calculated value of F, you

- increase the size of the numerator of the numerator—this means having treatments that are differentially effective
- decrease the size of the denominator of the numerator—this means having fewer treatment conditions which is not practical. Conversely the scientist cannot have too many conditions or otherwise the experiment becomes too complex. k will rarely exceed 6.
- decrease the size of the numerator of the denominator—this means that the participants in each treatment condition are uniformly responding to the treatment
- increase the size of the denominator of the denominator—this means increasing the number of participants in your sample

One rule of thumb captures the researcher's goal for maximizing the size of the F ratio to obtain statistical significance: *Heterogeneity between groups and homogeneity within groups.*

Critical Thinking Question: Can you link the two parts of this rule of thumb to the respective formulas in the F ratio and recognize how their calculations maximize the calculated value of F?

Example

A researcher was interested in the effect of practice on remembering a list of 15 words. Thirty college students were randomly divided into three groups of 10. The students in Group 1 were read the list three times and then asked to write down as many of the words as they could remember. Group 2 students were read the list twice and then asked to write down as many of the words as they could remember. Group 3 students were read the list once and then asked to write down as many of the words as they could remember. Each correctly recalled word was scored one point. The null hypothesis states that the means for the three groups will not differ from one another. That is, practice does not make a difference in remembering. The analysis of variance looks as follows:

X_1	X_2	X_3	X_1^2	X_2^2	X_3^2
10	4	5	100	16	25
9	6	2	81	36	4
8	4	8	64	16	64
7	8	7	49	64	49
12	6	10	144	36	100
10	4	2	100	16	4
11	5	1	121	25	1
12	7	4	144	49	16
9	10	3	81	100	9
8	11	4	64	121	16
96	65	46	948	479	288

$M_1 = 9.60$ $M_2 = 6.50$ $M_3 = 4.60$ $\Sigma X_t = 207$ $M_t = 6.90$ $\Sigma X_t^2 = 1715$

$$SS_1 = 948 - \frac{96^2}{10} = 26.40 \quad SS_2 = 479 - \frac{65^2}{10} = 56.50 \quad SS_3 = 288 - \frac{46^2}{10} = 76.40$$

$$F = \frac{\dfrac{(9.60 - 6.90)^2(10) + (6.50 - 6.90)^2(10) + (4.60 - 6.90)^2(10)}{3-1}}{\dfrac{26.40 + 56.50 + 76.40}{30-3}}$$

$$= \frac{\dfrac{SS_b}{df_n}}{\dfrac{SS_w}{df_w}} = \frac{\dfrac{127.40}{2}}{\dfrac{159.30}{27}} = \frac{63.70}{5.90} = 10.80$$

Chapter 4 presented the computational formula for the sum of squares. There is also a computational approach for computing the F. This approach means computing the SS_t and the SS_b and then subtracting SS_b from SS_t to obtain SS_w.

$$SS_t = \Sigma X_t^2 - \frac{(\Sigma X_t)^2}{N_t} = 1715 - \frac{207^2}{30} = 286.70 \quad \text{(Note that } \frac{(\Sigma X_t)^2}{N_t} \text{ is symbolized } C = 1428.30\text{)}$$

$$SS_b = \frac{(\Sigma X_1)^2}{n_1} + \frac{(\Sigma X_2)^2}{n_2} + \frac{(\Sigma X_3)^2}{n_3} - C = \frac{(96)^2}{10} + \frac{(65)^2}{10} + \frac{(46)^2}{10} - 1428.30 =$$

$$921.60 + 422.50 + 211.60 - 1428.30 = 127.40$$

$$SS_w = SS_t - SS_b = 286.70 - 127.40 = 159.30$$

$$MS_b = \frac{SS_b}{df_b} = \frac{127.40}{2} = 63.70 \qquad\qquad MS_w = \frac{SS_w}{df_w} = \frac{159.30}{27} = 5.90$$

$$F = \frac{MS_b}{MS_w} = \frac{63.70}{5.90} = 10.80$$

$F_{.01\,(2,27)} = 5.49$, $5.49 < 10.80$, so H_0 is rejected.

The Alternative Hypothesis

For the two-sample t test, there was just one alternative hypothesis, which could be expressed either directionally or nondirectionally. For research designs with more than two treatment conditions, there is more than one alternative hypothesis. For example, one alternative hypothesis is one group's mean is not equal to the second group's mean which in turn is not equal to the third group's mean, expressed symbolically as $H_1: \mu_1 \neq \mu_2 \neq \mu_3$. Another alternative hypothesis is one group's mean is not equal to the second and third groups' means which do not differ, expressed symbolically as $H_1: \mu_1 \neq \mu_2 = \mu_3$.

If the null hypothesis is rejected, then the challenge is to find out which alternative hypothesis is the correct one. This challenge requires using another statistical test called a post hoc test. Check your textbook to see whether it presents post hoc tests such as Tukey's Honestly Significant Difference test, the Tukey–Kramer test, the Newman–Keuls test, or the Scheffé test.

To answer the question of which H_1 is correct, we need to perform what is called a "post hoc test of multiple comparisons" to see what the correct alternative hypothesis is. There are a variety of these post hoc tests. One of the most preferred is called the Tukey's Honestly Significant Difference test because it does such a good job controlling for Type I error. There is a formula that computes the difference that two means need to meet or exceed in order for the means to be considered statistically significantly different from one another.

The formula: Tukey's HSD $= \alpha_{.05} \sqrt{\dfrac{MS_w}{n}} = 3.51 \sqrt{\dfrac{5.90}{10}} = 2.70$ (note that the $\alpha_{.05}$ comes from a

table such as the one at http://www.stat.duke.edu/courses/Spring98/sta110c/qtable.html). Now identify all pairs of the three means which differ from one another by at least 2.70. In this example, the mean of Group 1 (read three times group) is statistically significantly greater than the means of Groups 2 (read two times group) and 3 (read one time group), which do not differ from each other.

It is also necessary to compute effect size, which reflects the size of the influence of the treatment on the dependent variable. It is possible to have statistical significance (i.e., reject H_0) yet have a small effect.

Effect size $= \eta^2 = \dfrac{SS_b}{SS_t} = \dfrac{127.40}{286.70} = .44$, which represents a moderate effect (Sprinthall, 2012).

Assumptions to Use the Analysis of Variance

Assumptions for the analysis of variance (Sprinthall, 2012) include:
1. Treatment participants are randomly selected.
2. No member of one condition belongs to another condition (i.e., independent groups)
3. The population from which treatment groups are selected is normally distributed.
4. Data come from interval or ratio scales of measurement.
5. Within group variances of the treatment conditions should be similar
They are important to understand, for how they can affect the interpretation of the analysis of variance and indicating problems to avoid when you are collecting or analyzing your data.

Exercise

Using a highlighter, mark all of those sections of the passage below which you do not understand. You did this same exercise at the end of Chapter 1.

Rejecting the null hypothesis requires that the calculated value of F exceed the critical value of F coming from the table. Rejecting the null hypothesis indicates that the means of the three treatment conditions are different from one another. If the scientific hypothesis predicts that the treatment conditions will be different from one another, then the size of the calculated value of F should be maximized in order to reject the null hypothesis. From looking at the formula for the F ratio below, how can the size of the F be increased?

$$F = \frac{\dfrac{(M_1 - M_t)^2 n_1 + (M_2 - M_t)^2 n_2 + (M_3 - M_t)^2 n_3}{k-1}}{\dfrac{\Sigma(X_1 - M_1)^2 + \Sigma(X_2 - M_2)^2 + \Sigma(X_3 - M_3)^2}{N - k}}$$

Regard this formula for the F as being three fractions: the top fraction, the bottom fraction, and the overall fraction. Thus, there is the numerator of the numerator, the denominator of the numerator, the numerator of the denominator, and the denominator of the denominator.

Maximizing F occurs when there is heterogeneity between groups and homogeneity within groups. Thus, the F is increased if the size of the numerator of the numerator is increased and the size of the numerator of the denominator is decreased. In addition, F is increased as the sample size increase, which is reflected in the denominator of the denominator.

Stop! Compare the amount of material you underlined in the above passage with the amount of material you underlined for the same passage on p. 5 at the end of Chapter 1. The difference in the amount of material underlined is one assessment of your understanding of the one-way analysis of variance.

Problems

For Problems 1-3, there are three treatment groups each having 14 participants in them.

1. What are the degrees of freedom for the numerator of the F ratio?

2. What are the degrees of freedom for the denominator of the F ratio?

3. What is the critical value on the F distribution with 2 and 39 degrees of freedom and an $\alpha = .05$ (use http://www.statsoft.com/textbook/distribution-tables/#f05 or your textbook)?

4. True or false (if false, then correct): The greater the heterogeneity within groups and homogeneity between groups, the larger will be the calculated value of F.

In last chapter's example about the effectiveness of a new medication for reducing fever, the 25 patients receiving the new medication recorded a mean temperature change of $-1.46°$ with s of 1.33. The 25 patients receiving the placebo recorded a mean temperature change of $-.45°$ with an s of .94. A third condition is added with 25 patients receiving the regularly prescribed aspirin and the mean temperature change is -1.25 with an s of .53. Use the analysis of variance to answer the questions:

5. What is the sum of squares between groups?

6. What is the sum of squares within groups?

7. What is the mean square between groups?

8. What is the mean square within groups?

9. What is the calculated value of F?

10. Test the null hypothesis using $\alpha = .05$.

11. Researcher Problem: Children with cancer receive more emotional and informational support (EIS) during a week of summer camp than did children without cancer who did not attend summer camp. The camp experience was critical for receiving EIS, which is measured by Emotional/Informational Support Scale. Higher scores indicate less support.

Your team of researchers are asking whether there is something special about the summer camp experience for children with cancer or would chidren without cancer also demonstrate more EIS if they attended summer camp. So you perform a research study (based on Conrad, A. L., & Altmaier, E. M. (2009). Specialized summer camp for children with cancer: Social support and adjustment. *Journal of Pediatric Oncology Nursing, 26*, 150-157.) and have three groups each with 5 children:

Group 1 Children without cancer who are active in the summer but do not attend summer camp

Group 2 Children without cancer who are active in the summer and attend one week of summer camp.

Group 3 Children with cancer who are active in the summer and attend one week of summer camp for children with cancer.

The resulting EIS Scores for computer statistical analysis are:
Group 1: 40, 42, 43, 45, 46   Group 2: 33, 35, 36, 37, 46   Group 3: 31, 33, 35, 36, 39

1. What is the sum of squares between groups?
2. What is the sum of squares within groups?
3. What is the sum of squares total?
3. What are the degrees of freedom between groups?
4. What are the degrees of freedom within groups?
5. What is the mean square between groups?
6. What is the mean square within groups?
7. What is F?
8. Test the null hypothesis at the .05 level of significance.
9. What is the conclusion?

Chapter 11

The Dance of Correlation and Linear Regression

Chapters 2 through 6 covered the descriptive statistics for a group of scores where every member of the group had been measured once. In Chapters 7, 8, 9, and 10 you compared the differences between two or more groups of scores. In this chapter, you return to working with just one group of scores. However, *every member of the group is measured on two variables instead of just one*. Let me emphasize that again: Every member of the sample is measured on two variables instead of one. For each member, a pair of measurements is collected. What does it mean to ask "Are these two measurements related to each other?"

One goal of this chapter is to understand a relationship can exist between two measurements. As one measurement increases or decreases from one member of the sample to another, is the second measurement also changing its values? If yes, is the change in the same or opposite direction? A second goal is to present how one measurement can be used as the basis for *predicting* the second measurement if the relationship between the two variables is strong enough.

Positive Co–Relating as Line Dancing

Have you ever watched a group of people line dancing? Although some dancers take bigger steps or faster twirls than others, their movements are in synchrony. As one dancer steps to the right, the others also step to the right. As one dancer steps to the left, the others step to the left. These dancers are co–relating to each other in a positive direction.

Two measurements can similarly relate to each other. As one measurement increases, the other measurement increases. As one measurement decreases, the other measurement decreases. Think about the relationship between people's heights and weights. Generally, taller people are bigger and thus weigh more. Yes, some people are short and heavy, and some people are tall and thin. But overall, in a group of people, as height increases, weight tends to increase too.

The type of relationship when one measurement increases as the second measurement increases is called a positive correlation. A correlation is also positive when one measurement decreases as the second measurement decreases. For example, lower rainfall is associated with lower crop yield.

Negative Co–Relating as Slow Dancing

A couple with arms around each other glide along the dance floor to romantic music. Think about their synchronous movements as they face each other. As one partner moves to the right, which direction does the other partner move? To the *left*.

Similarly, a negative correlation occurs when one measurement made on a group increases when the other measurement decreases. Consider the relationship between job satisfaction and absences for a group of employees. As job satisfaction increases for this group of employees, what is likely to happen to the number of absences? The number will decrease. Conversely, as job satisfaction decreases, absences are likely to increase.

This correlation is negative. As one measurement increases, the other measurement decreases. Or as one measurement decreases, the other measurement increases.

No Co–Relating as Moshing

The final dance scene is a group of people moving randomly relative to each other as they do in the "disordered" state of the mosh pit (Silverberg, Bierbaum, Sethna, & Cohen, 2013). There is no synchronized pattern to the movements of the dancers. They move their bodies independently to the movements of the other dancers. Their dancing is not co–related with each other.

Similarly, two measurements may not correlate. The increase in one measurement is not matched with any systematic increase or decrease in the second measurement. When such is the case, the correlation is 0.

The Values for the Correlation

The values for the correlation range from –1 to +1. A correlation of –1 is a perfect negative correlation. As one person's measurement increases by a particular amount relative to another's measurement in the sample, the second measurement decreases by a particular amount.

A correlation of +1 is a perfect positive correlation. As one person's measurement increases by a particular amount relative to another's measurement in the sample, the second measurement increases by a particular amount.

A correlation of 0 means no correlation or no relationship between the two measurements. There is no pattern of increase or decrease between the two measurements within the sample.

As the correlation approaches 1 or –1, the relationship between the two measurements becomes stronger. Conversely, as the correlation approaches 0, the relationship becomes weaker. The strength of a correlation of .63 is the same as the strength of a correlation of –.63.

The Null and Alternative Hypotheses

The null hypothesis for the correlation is that there is no relationship. That is, the correlation will equal 0, H_0: $\rho = 0$. The Greek letter ρ is called rho, and is the correlation between the two measurements in the population. The sample correlation is symbolized with the letter r and estimates the population correlation.

The null hypothesis predicts that *no* relationship exists between the two measurements. The alternative hypothesis predicts that a relationship exists between the two measurements, that they are correlated. If the null hypothesis is rejected, then the researcher concludes that a relationship does exist.

CAUTION. Rejecting the null hypothesis means that there is a relationship between the two measurements. If the null is rejected, then the question becomes how strong or how weak is that relationship. The *strength* of the relationship is determined by the size of the correlation. It is possible to have a statistically significant relationship which is weak and possibly unimportant.

Introducing the Symbol Y for the Second Measurement

The descriptive and inferential statistical procedures covered previously in this book have analyzed only one measurement. This measurement has been symbolized with the letter X. The correlation requires that *two* measurements be made on each member of the data set. How is the second measurement symbolized? The second measurement is symbolized by the letter Y.

The use of X and Y should suggest to you that these measurements might be connected to the X and Y axes for graphing. You are correct if you are so thinking.

One important use for the correlation is a basis for prediction. Specifically, the stronger the correlational relationship between the two measurements, the better the measurement symbolized by the letter X predicts the measurement symbolized by the letter Y.

Critical Thinking Question: High school grade point average (GPA) and college GPA are correlated—are these two measurements positively correlated or negatively correlated? Positively, as high school GPA increases, college GPA also tends to increase.

Critical Thinking Question: Which one of the two measurements do you use to predict the other? You use high school GPA to predict college GPA.

The Formula for the Correlation

There are several formulas for the correlation. They look different but they are all algebraic variations of one another. As with the deviational and computational formulas for the sum of squares, some formulas are more understandable while other formulas are easier to work with.

For the correlation, the computational formula, the one easier to work with, is

$$r = \frac{\dfrac{\Sigma XY - (M_X)(M_Y)(N)}{N-1}}{s_X s_Y}$$

This formula introduces a new "character:" ΣXY, which is the *sum of the cross products*. Literally, for every pair of scores obtained from a member of the sample, the two scores are multiplied together. The resulting products are then added together.

There is another formula for the correlation which gives you a different "look" at how the correlation is computed. Even though they look different, both formulas are the same. The second formula is

$$r = \frac{\Sigma(X - M_X)(Y - M_Y)}{\sqrt{\Sigma(X - M_X)^2 \Sigma(Y - M_Y)^2}}$$

You can see the sum of squares for both X and Y in the denominator—they are multiplied together, and then the square root of the product is taken. Look at the numerator. It is similar to the sum of squares BUT instead of squaring a deviation score (i.e., multiplying it by itself), you multiply the deviation score for the X measurement by the corresponding deviation score for the Y measurement in the pair.

Testing the Null Hypothesis

Once the correlation is computed, it needs to be compared with the critical values to determine whether the null hypothesis is rejected or not rejected. The degrees of freedom for testing the null hypothesis for the correlation is N of the pairs of measurements - 2.

For critical values, go to http://www.gifted.uconn.edu/siegle/research/Correlation/corrchrt.htm. Compare the calculated value of r with the critical value for the particular level of α. If the calculated value is beyond the critical value, then the null hypothesis is rejected, and a relationship between the two measurements exists.

Understanding the Strength of the Correlation

For the correlation, rejecting the null hypothesis means only that a relationship exists between the two measurements but does not give any indication as to how strong that relationship is. To address this lack of information about relationship strength, Guilford (1956) assigned levels of strength of relationship to different sizes of correlations.

Correlations between -.20 and .20 indicate the relationship is slight, almost negligible. Negative correlations between -.20 and -.40 or positive correlations between .20 and .40 indicate a definite but small relationship. Negative correlations between -.40 and -.70 or positive correlations between .40 and .70 are moderate in strength, reflecting a substantial relationship. Negative correlations between -.70 and -.90 or positive correlations between .70 and .90 are high correlations, reflecting a marked relationship. Negative correlations between -.90 and -1.00 or positive correlations between .90 and 1.00 are very high correlations indicating a very dependable relationship.

Assumptions for Using the r

The assumptions for performing the correlation (Sprinthall, 2012) include:
1. The sample is randomly selected from the population.
2. Both measurements are normally distributed.
3. At least interval level data are used.
4. Variance of both scores are similar.
5. Linear relationship exists between X and Y.

Graphing

All the graphing done so far in the course has been for a univariate distribution. Univariate means that one measurement has been obtained from every member of the sample or population.

For example, you convert a data set into a frequency polygon so you can "eyeball" the data. You plot your scores on the X axis and frequency or relative frequency on the Y axis. Graphing for the t test or analysis of variance places the scores on the Y axis and the groups (e.g., experimental and control) on the X axis.

For the correlation, *two* measurements have been obtained from every member of the data set. One of the measurements is graphed on the X axis, and the other measurement is graphed on the Y axis. Which goes where?

When a correlation is graphed, the measurement *being used to predict* is graphed on the X axis and is symbolized as X. The measurement *being predicted* is graphed on the Y axis and is symbolized as Y. Again remember $Y = f(X)$. Y is a function of X, which means that the value of Y is dependent on the value of X. The Y measurement acts as the dependent variable and is graphed on the Y axis.

The distribution for the correlation is called a *bivariate* distribution because it contains two measurements. Each point on the graph represents the pair of (X, Y) scores obtained from each member of the data set. This graph is also called a *scatter plot*.

The shape of the pattern of points in the scatter plot is important. As the correlation between two measurements approaches either –1 or 1, the points increasingly line up. When the correlation is –1 or 1, the points form a straight line. Conversely, as the correlation gets closer to 0, the scatter of the dots becomes less linear. The correlation of 0 means the points are a random scatter.

Linear Regression—The Crystal Ball of Statistics

Once a correlation has been computed, the relationship between two measures is established. If this relationship is strong enough (i.e., close to -1 or 1), then one measurement can be used to predict the second measurement. The process of developing the "crystal ball" for prediction is called *regression*.

Simple linear regression describes the situation when you have just one predictor variable. Multiple linear regression, on the other hand, is when you have two or more predictor variables. This section presents simple regression only.

The *crystal ball* is the formula for the line that passes the closest through all of the points in the scatter plot. This *line of best fit* is called the *regression line*, and the formula for the line is $\hat{Y} = a + bX$. Let's take this formula apart:

- $\hat{Y}$ is called "Y hat" and is the predicted value of Y computed using the corresponding X score
- a is called the Y intercept and is the point at which the regression line crosses the Y axis
- b is the slope of the regression line
- X is the value used to predict Y

When each X has been included in the formula and its corresponding $\hat{Y}$ computed, then the resulting pairs of $\left(X, \hat{Y}\right)$ values are plotted on the bivariate distribution. These points connect to form a straight line called the regression line. This regression line is the *line of best fit* through the scatter plot of the (X, Y) dots.

Three Values of Y for Every Value of X

Every member of the data set is measured twice—one for the X measurement and one for the Y measurement. Finding the regression line requires computing $\hat{Y}$ for every X. There is a Y and a $\hat{Y}$ for every X measurement.

There is a third value of Y for each X, and this is the mean of Y, which is symbolized as M_Y. When all the (X, M_Y) values are plotted on the bivariate distribution, the points form a line which is parallel to the abscissa because the same value of M_Y exists for all values of X.

The Tension of Prediction

Which is the better predictor for Y:
- M_Y or the mean of Y which is computed based solely on the Y scores, or
- $\hat{Y}$, the predicted value of Y computed based on the additional information represented by X

To answer this question, it is necessary to see what proportion of the total sum of squares of Y is accounted for or explained by X. To find out this proportion, you split the sum of squares of Y into two pieces—the sum of squares of Y not explained by X and the sum of squares of Y explained by X.

Critical Thinking Question: What is the formula for the sum of squares of Y? SS_Y = sum of squares of $Y = \Sigma(Y - M_Y)^2$.

This deviational formula for the sum of squares is based on the deviation score $Y - M_Y$. Consider Y as your dorm room, apartment, or house. Consider M_Y as the location of the classroom for your statistics course. Now select a landmark between where you live and where you have class. The total distance between where you live and where you have class can now be split into the distance between where you live and the landmark and the distance between the landmark and where you have class.

For regression, the landmark is $\hat{Y}$. The total distance between the raw score Y and the mean M_Y can be split into the resulting differences of $Y - \hat{Y}$ and $\hat{Y} - M_Y$. Once these differences are calculated for each $\hat{Y}$, you can produce the following two sums of squares:

- $SS_{Y \text{ not explained by } X} = SS_{\text{errors of prediction}} = \Sigma\left(Y - \hat{Y}\right)^2$
- $SS_{Y \text{ explained by } X} = \Sigma\left(\hat{Y} - M_Y\right)^2$

These two sums of squares add together to equal the sum of squares for Y:

$$SS_Y = \Sigma(Y - M_Y)^2 = \Sigma(Y - \hat{Y})^2 + \Sigma(\hat{Y} - M_Y)^2$$

The Coefficient of Determination

The coefficient of determination is the proportion of the sum of squares of Y explained by X:

$$\text{coefficient of determination} = \frac{SS_{Y \text{ explained by } X}}{SS_Y} = \frac{\Sigma(\hat{Y} - M_Y)^2}{\Sigma(Y - M_Y)^2}$$

The coefficient of determination is also computed as r^2.

The Standard Error of Estimate

We can compute the spread of points around the regression line of the bivariate distribution as we computed the spread of scores around the mean of the univariate distribution.

Critical Thinking Question: What are the three steps for computing the standard deviation?
1. Compute the sum of squares.
2. Divide the sum of squares by the degrees of freedom to compute the variance.
3. Take the square root of the variance to find the standard deviation.

For the univariate distribution, the mean is the referent for computing the standard deviation. For the bivariate distribution, the regression line is the referent for computing the standard error of estimate. Deviation is defined as the difference between the Y and the $\hat{Y}$ values. These differences are then squared and summed, the sum is divided by the degrees of freedom $(N - 2)$, and the square root of the quotient is computed. Thus,

$$SE_{est} = \sqrt{\frac{\Sigma(Y - \hat{Y})^2}{N - 2}}$$

Another way of computing the standard error of estimate is $SE_{est} = s_Y \sqrt{\frac{N(1 - r^2)}{N - 2}}$

Example

You are a personnel director who wants to hire only those people who are going to become the best workers. Is there a way of determining how good a worker will be *before* he or she is hired? Given the manufacturing nature of your company, the workers need good eye–hand coordination because they assemble the product as the pieces pass the employee on a conveyor belt. If people with the best visual motor integration (VMI), or eye-hand coordination, are the most productive workers, then you can use a test of VMI to help you hire the best people. First, though, you must determine whether a relationship exists between VMI scores and job performance. You administer a test of VMI to the next 10 hirees, and three months into the probation period, you collect their first work ratings. Your data are tabulated below (note: higher VMI scores and Work Ratings indicate better performance):

Subject ID#	X (VMI scores)	Y (Work Ratings)
1	11	6
2	12	7
3	9	5
4	19	9
5	15	6
6	6	4
7	1	2
8	8	4
9	18	8
10	14	7

1. Compute the mean of Y. $M_Y = 5.8$.

2. For each subject, multiply the X and Y pair of scores. For example, the product of the two scores for subject #1 is 66. Add the 10 products together, and you have computed ΣXY, 756.

3. Make a scatter plot of your data by graphing the data points on the graph paper below. Label both axes.

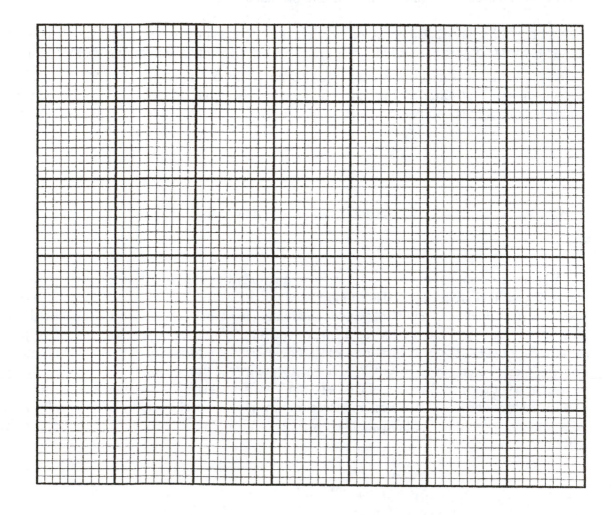

Critical Thinking Question: Why are the VMI scores assigned X and the Work Ratings assigned Y? Because VMI scores are used to predict Work Ratings.

4. Add the 10 points for the (X, M_Y) coordinates, and connect the points. Hint: You will end up with a line parallel to the abscissa. Contrast the line with the scatter of the points. How closely does the line follow the pattern of the points?

5. Find the s of X and the s of Y. The s of X is 5.53 and the s of Y is 2.10.

6. Now compute $\quad r = \dfrac{\dfrac{\Sigma XY - (M_X)(M_Y)(N)}{N-1}}{s_X s_Y} = \dfrac{\dfrac{756 - (11.30)(5.80)(10)}{9}}{(5.53)(2.10)} = .96$

7. The formula for the regression line is $\hat{Y} = a + bX$. The formulas for a or b are $b = \dfrac{r(s_Y)}{s_X}$ and $a = M_Y - b(M_X)$. Compute a and b. $a = 1.73$ and $b = .36$.

8. Using the regression line formula, compute all 10 values of $\hat{Y}$, plot the 10 ($X, \hat{Y}$) points on the graph on the previous page, and connect the points. You now have two lines drawn on your graph—the M_Y and the $\hat{Y}$ line. The $\hat{Y}$ line is called the regression line. Which of these two lines better follows the pattern of your (X, Y) points? The regression line.

Remember that the formula for the regression line becomes the crystal ball. If you want to hire someone who will have a work rating of at least 6 or better (i.e., above the mean) after the first three months on the job, then you will tend to advance to subsequent levels of the hiring process those applicants who score 11 or better, unless the applicant has other qualifications like good recommendations or broad job experience.

CAUTION: In 1993, the Target Department Store settled a law suit filed on behalf of 2,500 job applicants for security guard positions who completed the Rodgers Condensed California Personality Inventory (CPI)–Minnesota Multiphasic Personality Inventory (MMPI) instrument. The applicants alleged that the questions were not predictive of their performance as a security guard, and the court agreed.

9. Now compute SS_Y, $SS_{Y \text{ not explained by } X}$, $SS_{Y \text{ explained by } X}$.

- $SS_Y = \Sigma(Y - M_Y) = 39.60$

- $SS_{Y \text{ not explained by } X} = \Sigma(Y - \hat{Y})^2 = 2.94$

- $SS_{Y \text{ explained by } X} = \Sigma(\hat{Y} - M_Y) = 36.66$

10. Finally, compute the coefficient of determination:

$$r^2 = \frac{\Sigma(\hat{Y} - M_Y)^2}{\Sigma(Y - M_Y)^2} = \frac{36.66}{39.60} = .93$$

The coefficient of determination is translated as meaning that 93% of the variability of the work ratings is explained by the VMI scores. The correlation between VMI scores and work ratings is very strong, so VMI scores are a good predictor of work ratings.

Putting It All Together

- One dot on the bivariate scatter plot corresponds to one pair of (X, Y) scores.
- As the correlation approaches -1 or 1, the stronger the relationship between the two measurements.
- As the correlation approaches -1 or 1, the more linear will be the pattern of the (X, Y) dots
- As the correlation approaches 0, the more randomly scattered will be the (X, Y) dots
- The stronger the relationship, the more the variability of Y is explained by X, the larger the coefficient of determination, and the better X predicts Y.
- If the correlation equals 1 or -1, then $\Sigma(Y - M_Y)^2 = \Sigma(\hat{Y} - M_Y)^2 , \Sigma(Y - \hat{Y})^2 = 0$, and the standard error of estimate $= 0$
- The mean is to the univariate distribution as the regression line is to the bivariate distribution.

Problems

1. Are high school GPA and college GPA positively or negatively correlated?

2. When graphing these two variables, which is placed on the X axis and which is placed on the Y axis?

3. As the correlation becomes stronger, X becomes a _____ (better, poorer) predictor of Y.

4. Compute the correlation for the following five pairs of scores:

X	Y
1	1
2	2
3	3
4	4
5	5

Use the following five pairs of scores to answer questions 5 through 10:

X	Y
1	5
2	4

3   3
4   2
5   1

5. What is the correlation?

6. What is the formula for the regression line?

7. What is the sum of squares for Y?

8. What is the sum of squares for Y not explained by X?

9. What is the sum of squares for Y explained by X?

10. Compute the coefficient of determination two different ways?

11. The mean is to the univariate distribution as the _____ is to the bivariate distribution.

12. Researcher Problem: The local Chamber of Commerce wants to be supportive of the university students and their economic needs. Chamber members recognize that many students work, and they are concerned that students might need to work to afford rising gas prices. The more money spent on gas, the less money spent on goods and services with local businesses.

The Chamber of Commerce hires you to answer two questions:

1. Is there a relationship between the number of hours worked per week and the amount of money spent per month on gas?
2. How strong is that relationship?

You identify a sample of 33 university students and ask them two questions:
• How many hours do you work each week?
• How much money do you spend each month on gas?

The descriptive data are as follows with WORKHRS symbolizing the number of hours worked each week and GAS$MTH symbolizing the amount of money spent each month on gas:

Descriptive Statistics

	Mean	Std. Deviation	N
WORKHRS	16.97	15.226	33
GAS$MTH	111.36	85.826	33

You also compute the sum of the cross products (see below for data), which is 83,065.

What is the correlation?

What is the coefficient of determination?

What are your answers to the two questions for the Chamber of Commerce?

Here are the data if you want to analyze them with SPSS or some other statistical software:

WORKHRS	GAS$MTH
11	100
40	120
40	100
10	175
20	30
7	100
20	80
29	40
0	0
20	60
0	100
20	200
0	15
20	60
0	65
20	120
15	75
4	60
0	100
0	80
55	375
0	50
20	50
30	75
0	250
30	100
20	250
40	300
20	200
45	175
0	30
8	100
16	40

Chapter 12

The Chi-Square Test

Working with Nominal Data

Chapter 2 presented the four scales of measurement: nominal, ordinal, interval, and ratio. The nominal scale of measurement included numbers used as names (e.g., football player wearing 45 on his jersey, runner wearing 21 on her jersey) or category memberships (e.g., Democrats coded as 1 and Republicans coded as 2). Can nominal data be statistically analyzed? The answer is yes. Here are two different problems which the chi square can help answer.

Problem 1: Goodness of Fit Test. Statistical State University (SSU) has four sororities on campus: $\Omega\Omega\Omega$, $\beta\beta\beta$, $\Phi\Phi\Phi$, and $\Psi\Psi\Psi$. Is one of these sororities more respected by the women of SSU than the other three? The researcher selects a random sample of 100 women and asks them to indicate which one of the four sororities they respected the most. These data are called the *observed frequencies*. In contrast, the *expected frequencies* are based on what would be expected if each sorority was equally respected by the 100 women. These expected frequencies are called the frequencies expected by chance. Expected frequencies can also be the result of theoretical prediction.

Thus, by chance, the expected frequencies for the 100 women would be 25 women most respecting $\Omega\Omega\Omega$, 25 women most respecting $\beta\beta\beta$, 25 women most respecting $\Phi\Phi\Phi$, and 25 women most respecting $\Psi\Psi\Psi$. The chi square formula compares the observed frequencies with the expected frequencies.

Problem 2: Test for Independence. The Dean of Statistical State University was interested in whether the men and women students majored more in arts and sciences, business, or education. A random sample of 100 men and 100 women was selected. The observed frequencies for the men were 35 men in arts and sciences, 20 men in education, and 45 men in business. The observed frequencies for the women were 29 women in arts and sciences, 48 women in education, and 23 women in business. Computing the expected frequencies is shown in the **Solving Problem 2** section on page 101.

Formula, Null and Alternative Hypotheses, and Degrees of Freedom for the Chi Square

The formula for the chi square compares the observed frequencies, symbolized as f_o, and the expected frequencies, symbolized as f_e. This comparison occurs through subtracting f_e from f_o Yet again, another difference, $f_o - f_e$, is at the heart of a statistical procedure.

Critical Thinking Question: What other differences encountered in previous chapters are important in statistics? Deviation score and sampling error

The formula for the chi square is:

$$\chi^2 = \Sigma \frac{(f_o - f_e)^2}{f_e}$$

The null hypothesis is that the observed frequencies and expected frequencies equal one another; $H_0: f_o = f_e$. The alternative hypothesis is that the observed frequencies and expected frequencies do not equal one another; $H_1: f_o \neq f_e$.

The degrees of freedom for Problem 1 is the number of categories - 1 = 4 - 1 = 3. The degrees of freedom for Problem 2 is the (number of rows - 1) x (the number of columns - 1) = 1 x 2 = 2.

Solving Problem 1

The observed frequencies for respecting sororities were 40 women most respecting ΩΩΩ, 25 women most respecting βββ, 15 women most respecting ΦΦΦ, and 20 women most respecting ΨΨΨ. The expected frequencies for respecting sororities, based on chance, were 25 women most respecting ΩΩΩ, 25 women most respecting βββ, 25 women most respecting ΦΦΦ, and 25 women most respecting ΨΨΨ. Here are the tabulated data and computations:

observed frequency (f_o)	40	25	15	20
expected frequency (f_e)	25	25	25	25
$f_o - f_e$	15	0	-10	-5
$(f_o - f_e)^2$	225	0	100	25
$\dfrac{(f_o - f_e)^2}{f_e}$	9	0	4	1

$$\chi^2 = \Sigma \frac{(f_o - f_e)^2}{f_e} = 9 + 0 + 4 + 1 = 14$$

Is the chi square value of 14 statistically significant? To answer this question, two additional pieces of information are necessary: how many degrees of freedom are there and what is the critical value? The degrees of freedom are the number of categories - 1, so 4 - 1 is 3 degrees of freedom.

The critical value for χ^2 is found in the chi-square table at http://www.statsoft.com/textbook/ distribution-tables/#chi or in the chi-square table in the back of a statistics textbook. For 3 degrees of freedom, the critical value for $\alpha = .05$ is 7.81. The critical value for $\alpha = .01$ is 11.34. The calculated value of 14 is greater than the critical values for either level of alpha, meaning that the null hypothesis is rejected. More women respect ΩΩΩ sorority than the other three sororities.

Solving Problem 2

For Problem 2, a random sample of 100 men and 100 women was selected. The observed frequencies for the men were 35 men in arts and sciences, 20 men in education, and 45 men in business. The observed frequencies for the women were 29 women in arts and sciences, 48 women in education, and 23 women in business. The data for Problem 2 are represented below in what is called a 2 x 3 contingency table, which has six cells:

	Arts and Sciences	Education	Business	
Men	A 35	B 20	C 45	100
Women	D 29	E 48	F 23	100
	64	68	68	200

Besides the observed frequencies inside each cell, the table has two other important features. First, in each cell with the observed frequency is a letter. The letter identifies that cell and will be useful for you when understanding the table below. The second feature is the numbers in the margins of the table. The row margins, which occur at the end of each row, contain 100 for 100 men and 100 for 100 women. The column margins contain 64 for 64 Arts and Science majors, 68 for 68 Education majors, and 68 for 68 business majors.

Computing expected frequencies. The numbers inside the table are the observed frequencies. The numbers in the row and column margins are used to compute the expected frequencies. For example, look at Cell A, which contains 35 men majoring in arts and sciences.   To find the expected frequency, multiply Cell A's row marginal by its column marginal and divide the product by 200. Thus, the expected frequency for the number of men majoring in arts and sciences is $\frac{100 \times 64}{200} = 32$. For Cell E, the expected frequency is also $\frac{100 \times 68}{200} = 34$. Here is the entire table:

	A	B	C	D	E	F
observed frequency (f_o)	35	20	45	29	48	23
expected frequency (f_e)	32	34	34	32	34	34
$f_o - f_e$	3	-14	11	-3	14	-11
$(f_o - f_e)^2$	9	196	121	9	196	121
$\dfrac{(f_o - f_e)^2}{f_e}$	.28	5.76	3.56	.28	5.76	3.56

$$\chi^2 = \Sigma \frac{(f_o - f_e)^2}{f_e} = .28 + 5.76 + 3.56 + .28 + 5.76 + 3.56 = 19.20$$

The critical value for χ^2 is found in the chi-square table at http://www.statsoft.com/textbook/

distribution-tables/#chi or in a chi-square table in the back of a statistics textbook. For 2 degrees of freedom, the critical value for α = .05 is 5.99. The critical value for α = .01 is 9.21.   The calculated value of 19.20 is greater than the critical values for either level of alpha, meaning that the null hypothesis is rejected.   There is a statistically significant difference between men and women and what they major in at SSU.

Assumptions for Using the Chi Square

There are four assumptions for performing the correlation (Sprinthall, 2012):
1. The data are nominal.
2. The sample is randomly selected from the population
3. A member of the data set can appear on only one category or one cell.
4. Expected frequency values should be 5 or higher.

Problems

1. Data from what scale of measurement are used in the chi-square test?

2. What are the names of the two different chi-square tests?

3. How do you compute the degrees of freedom for each of these two tests?

4. How many cells are in a 3 x 3 contingency table?

5. What is the critical value of χ^2 when there are 5 degrees of freedom and an α =.05 ?

6. What is the critical value of χ^2 when there is 1 degree of freedom and an α =.01 ?

7. For the Goodness of Fit test, how are the expected frequencies computed?

8. For the Test of Independence, how are the expected frequencies computed?

9. A candidate for mayor claims that 65% of the voters will vote for her. A random sample of 300 voters indicates that 185 expected to vote for the mayor. Use the chi square test to determine whether the candidate's prediction is true for the sample data?

10. A researcher is interested in knowing whether hospitals in urban areas performed more Caesarean sections than hospitals in rural areas.   For a random sample of 50 urban women, urban hospitals had delivered 40 via natural childbirth and 10 via Caesarean sections. For a random sample of 50 rural women, rural hospitals had delivered 48 via natural childbirth and 2 via Caesarean section. Use the chi square test to test the null hypothesis.

References

Centers for Disease Control and Prevention. CDC growth charts. http://www.cdc.gov/growthcharts/cdc_charts.htm. Available April 30, 2013.

Cohen, J. (1994). The earth is round ($p < .05$). *American Psychologist, 49*, 997–1003.

Feldman, R. S. (2011). *Development across the life span* (6th ed). Upper Saddle River, NJ: Prentice Hall.

Hagen, R. L. (1997). In praise of the null hypothesis statistical test. *American Psychologist, 52*, 15–24.

Kranzler, J. H. (2011). *Statistics for the terrified* (5th ed.). Englewood Cliffs, NJ: Prentice Hall.

Krueger, J. (2001). Null hypothesis significance testing: On the survival of a flawed method. *American Psychologist, 56*, 16-26.

Larsen, J. K., Bendsen, B. B., Foldager, L., & Munk-Jørgensen, P. (2010). Prematurity and low birth weight as risk factors for the development of affective disorder, especially depression and schizophrenia: A register study. *Acta Neuropsychiatrica, 22*, 284-291.

Lovejoy, E. P. (1975). *Statistics for math haters*. New York: Harper & Row.

Phillips, J. L. Jr. (2000). *How to think about statistics* (6th ed.). New York: W. H. Freeman and Co.

Rowntree, D. (1981). *Statistics without tears: A primer for non–mathematicians*. New York: Charles Scribner's Sons.

Silverberg, J. L., Bierbaum, M., Sethna, J. P., & Cohen, I. (2013). Collective motion of moshers at heavy metal concerts. arXiv: 1302.1886 v1 [physics.soc-ph]. Retrieved from http://arxiv.org/abs/1302.1886.

Sprinthall, R. C. (2012). *Basic statistical analysis* (9th ed.). Boston: Allyn and Bacon.

Stevens, S. S. (1951). *Handbook of experimental psychology*. New York: John Wiley & Sons.

Weaver, K. A., Drone, C. M., Varela, E., & Heskett, C. (1996, August). *Variables influencing introductory statistics achievement.* Paper presented at the annual convention of the American Psychological Association, Toronto, Canada.

Weaver, K. A. (1992). Elaborating selected statistical concepts with common experience. *Teaching of Psychology, 19*, 178-179.

Weaver, K. A. (1999). The statistically marvelous medical growth chart: Tool for teaching variability. *Teaching of Psychology, 26*, 284-286.

Answers

Chapter 1

1. 10.49
2. 470,558.56
3. 1876.51
4. 69.80
5. −12.63
6. 68.40
7. −.79
8. −.24
9. 47.17
10. −28.56
11. −41.08
12. 21.75
13. 12.34
14. 66.33
15. .30
16. .95
17. 332.33
18. 1,674,436
19. .22
20. 48.72
21. 17.02
22. .86
23. false
24. True
25. false, N–D is 0
26. true
27. true
28. true
29. true
30. false
31. Integers: 8 and 121.00
32. .25
33. 6.83
34. 754.94
35. 18.94
36. 25643.34
37. true
38. false
39. true
40. true
41. false
42. false

Chapter 2

1. The scores in a data set have been put into numerical order.
2. Descriptive statistics describe a group. Inferential statistics attempts to describe a population based on information obtained from the sample.
3. False, 65 is not a member.
4. False, Fahrenheit is an interval not a ratio measure of temperature.
5. False, test scores are an interval not a ratio measure of intelligence.
6. False, test scores are an interval not a ratio measure of achievement or knowledge.
7. True, the number of points belongs to the ratio scale of measurement. The 0 is meaningful.
8. true
9. #88–nominal, 5 mile–ratio, first place–ordinal, 26.50 minutes–ratio
10. ratio

Chapter 3

1. a. people's salaries are *positively* skewed
   b. housing costs are *positively* skewed
   c. the distribution of responses resulting from men being asked "If you received $1 for every sexist thought you had in the past year, how much richer would you be today?" are *positively skewed*.
2. negative skewness indicating that most students did well
3. The number of intervals is a number between 10 and 15. The interval size is found by dividing the number intervals into the range of 163 and rounding the answer up. For example, if I wanted 10 intervals, then the size of each of those intervals is 17 because $10\overline{)163} = 16.3$, which is 17 when rounded up to the integer.

4. a. The median is the score located in the $\frac{15+1}{2} = 8$ position. Counting from either 38 or from 89, the median is 41.

    b. The distribution is skewed positively.

5. c. to the left of

6. If two–thirds of the sample have scores less than the mean, then the distribution is **positively** skewed. Positive skewness means that the distribution has more high scores than one expects with a normal distribution. The high scores inflate the mean, "pulling" it to the far right side of the distribution. Consequently, most of the scores will be below or less than the mean.

7. Sample is to population as **e. b and d** are to the even numbers less than 100.
    a. 1, 2, 90 b. 2, 4, 18, 92 c. 6, 28, 100 d. 96, 98 e. b,d f. b, c, d g. b, c h. c, d
    Only samples in selections **b** and **d** are subsets of the even numbers less than 100.

8. b. median

For questions 9–12 use the distribution 5, 9, 11, 11, 12, 14, 15, 17, 18, 98, 100:

9. $\Sigma X = 310$, $N = 11$, $M = \dfrac{\Sigma X}{N} = \dfrac{310}{11} = 28.181818 = 28.18$

10. If the N is odd, the median is the score located in the $\dfrac{N+1}{2}$ position. N equals 11, so the median is the score located in the $\dfrac{11+1}{2}$ or $\dfrac{12}{2}$ or 6th position. Starting with either the lowest score or the highest score and counting 6, the median is 14 as follows:
    5, 9, 11, 11, 12, (14), 15, 17, 18, 98, 100.

11. The mode is 11.

12. The mode is less than the median which is less than the mean, skewing the distribution positively. 98 and 100 are much larger than the other scores, distorting the distribution to the far right.

13. False, social security number belongs to the nominal scale.

Chapter 4

1. 0, if all the scores are the same, then there is no deviation from the mean.

2. true

3. variability

4. —

5. mean

6. cluster

7. $\Sigma X = 425$, $M = 85$

8. $SS = \Sigma x^2 = 250$ OR $SS = \Sigma X^2 - \dfrac{(\Sigma X)^2}{N} = 36{,}375 - \dfrac{(425)^2}{5} = 250$

9. $V = \dfrac{SS}{N} = \dfrac{250}{5} = 50$

10. $SD = \sqrt{V} = \sqrt{50} = 7.07$

11. The standard deviation of the first distribution is 1.72, the standard deviation of the second distribution is 5.16. The first distribution is more homogeneous.
12. $V = SD^2$, so the variance of Section A is $6^2 = 36$, and the variance for Section B is $8^2 = 64$. Section B's distribution is more platykurtic because its SD is larger.

Chapter 5

1. Variance, sum of squares, and standard deviation can only be positive.
2. $z = \dfrac{17 - 21}{3} = \dfrac{-4}{3} = -1.33$
3. $50\% - 40.82\% = 9.18\%$
4. $50\% + 40.82\% = 90.82\%$
5. One–half of 200 or 100
6. $200 \times .9082 = 181.64$
7. $200 \times .4082 = 81.64$
8. $z = \dfrac{23 - 21}{3} = \dfrac{2}{3} = .67$, so 24.86%
9. $z = \dfrac{20 - 21}{3} = \dfrac{-1}{3} = -.33$
10. $z = \dfrac{15 - 21}{3} = \dfrac{-6}{3} = -2; z = \dfrac{26 - 21}{3} = \dfrac{5}{3} = 1.67$; 15 is farther from the mean.

Chapter 6

1. $z = \dfrac{85 - 100}{15} = \dfrac{-15}{15} = -1$. The score of 85 is located 1 standard deviation to the left of the mean. To find the probability of selecting a score which is less than 85, subtract .3413 from .5000 and the answer is .1587, which is .16 when rounded to the hundredth.
2. The most extreme 5% of the distribution's total area is far in the tails, beginning at ±1.96 standard deviations. To find the raw scores, sse the $X = (z)(SD) + M$ formula. Thus, (±1.96)(15) + 100 = 70.60 and 129.40.
3. true
4. —
5. ±1.65 are the z scores dividing the distribution between the innermost or middlemost 90% of the area and the outermost 10% of the area.
6. Apply the ADD–OR rule: $\dfrac{4}{52} + \dfrac{4}{52} = \dfrac{8}{52} = .15$
7. Apply the MULT–AND rule: $.50 \times .50 \times .50 \times .50 = .06$
8. true
9. true The smallest value for probability and the standard deviation when all the scores in the distribution are the same both equal 0
10. False $1 \neq 0$
11. —

Chapter 7

1. That the sample is not representative of the population.
2. The sampling distribution of the mean becomes more leptokurtic because the samples are becoming more similar to the population and the sample means are thus becoming more similar (i.e., less variable) to the population mean.
3. reject the null hypothesis
4. $\sigma_M = \dfrac{\sigma}{\sqrt{N}} = \dfrac{15}{\sqrt{48}} = 2.17$
5. $\sigma_M = \dfrac{\sigma}{\sqrt{N}} = \dfrac{15}{\sqrt{36}} = 2.50$ $2.17 < 2.50$ so the distribution becomes more platykurtic.
6. The population mean is to the population distribution as **c.** μ is to the sampling distribution of the means.
7. .05
8. Deviation score is to **SAMPLE** distribution as **SAMPLING ERROR** is to sampling distribution of the means.   **d. sample, sampling error**
9. population mean or mean of the sampling distribution of the means
10. σ
11. An $\alpha = .01$ means that the alpha area would equal the extreme 1% of the area under the normal curve. Given the symmetry of the normal curve, this means that .5% of the area would be located in the extreme left tail and .5% of the area would be located in the extreme right tail. Thus, 49.50% of the area would be located from the boundary of the alpha area to the mean of the distribution. At http://www.statsoft.com/textbook/distribution-tables/#z, look inside the table for .4950. You will find .4949 and .4951. Using the same thinking as in rounding, because .4950 is halfway between the two values, round up to .4951. What z score corresponds to .4951? It is 2.58, thus the boundaries into the alpha area when $\alpha = .01$ are ± 2.58 standard errors of the mean.
12. $\sigma_M = \dfrac{15}{\sqrt{25}} = 3$; $z = \dfrac{105 - 100}{3} = 1.67$; critical values are ± 1.96. The null hypothesis is not rejected; the sample is representative of the population; the manuscript was published.

Chapter 8

1. True. As sample size increases, the t distribution becomes more like the z distribution. With less and less of the area in the tails, the boundaries into the alpha area move closer to the sampling distribution's mean.
2. 27
3. ± 2.042
4. estimated standard errors of the mean
5. closer to
6. more
7. false, alpha is five times larger when it is .05 instead of .01
   For 8, 9, and 10: $N = 5$, $\Sigma X = 523$, $M = 104.60$, $\Sigma X^2 = 55{,}273$, $SS = 567.20$
8. $s = 11.91$
9. $SE_M = 5.33$

$$t = \frac{104.60 - 108}{5.33} = \frac{-3.40}{5.33} = -.64$$

10. $t_{4;.05} = \pm 2.78$

    -.64 is not located in the alpha area so the null hypothesis is not rejected; thus the sample mean of 104.60 is representative of the population

11. inequality

12. ± 1.96

13. Researcher Problem

X	X^2	X-M	$(X-M)^2$
99.00	9801.00	0.80	0.64
97.00	9409.00	-1.20	1.44
98.60	9721.96	0.40	0.16
97.70	9545.29	-0.50	0.25
98.30	9662.89	0.10	0.01
98.00	9604.00	-0.20	0.04
98.60	9721.96	0.40	0.16
98.80	9761.44	0.60	0.36
97.90	9584.41	-0.30	0.09
98.10	9623.61	-0.10	0.01
982.00	96435.56	0.00	3.16

$M = 98.2°$

$\Sigma x^2 = 3.16$ (deviational approach for computing sum of squares)

$\Sigma X^2 - (\Sigma X)^2/N = 96435.56 - 982^2/10 = 96435.56 - 964324/10 = 96435.56 - 96432.40 = 3.16$
        (computational approach for computing sum of squares)

$s = \sqrt{SS/N-1} = \sqrt{3.16/9} = .59$

$SE_M = s/\sqrt{10} = .59255/3.16227 = .18738$

$t = M-\mu/SE_M = 98.20-98.60/.18738 = -.40/.18738 = -2.13$

Critical values of t for 9 degrees of freedom and $\alpha = .05$ are ± 2.262

Fail to reject H_0

The body temperature of the sample of 10 adults does not differ from 98.6° so the research study can begin

Chapter 9

1. $s_E = \sqrt{\dfrac{250}{25}} = \sqrt{10} = 3.16,\ SE_{M_E} = \dfrac{3.16}{\sqrt{26}} = .62$

   $s_C = \sqrt{\dfrac{300}{25}} = \sqrt{12} = 3.46,\ SE_{M_C} = \dfrac{3.46}{\sqrt{26}} = .68$

   $SE_D = \sqrt{.62^2 + .68^2} = \sqrt{.8468} = .92$

2. $t = \dfrac{20 - 24}{.92} = \dfrac{-4}{.92} = -4.35$

3. With 50 degrees of freedom, an alpha of .05, and a one–tailed test, the critical value of t is −1.677.

4. Reject the null hypothesis. Note that the null is rejected with an alpha of .01 as well.

5. Directional alternative hypothesis.

6. Null hypothesis

7. α and σ increase the likelihood of rejecting the null hypothesis

8. Alpha is the probability of incorrectly rejecting a true null hypothesis.

9. Alpha error is considered more serious because it indicates that the treatment is effective when it is not. The potential for harm is greater when changing to use an ineffective treatment than it is to keep on doing what you normally do and miss an opportunity to improve.

10. Use the one–tailed t test when a directional alternative hypothesis has been made.

11. Nondirectional, directional

12. I, II

13. sampling distribution of differences

14. Two answers: Either the standard deviation of the population or the standard error of the mean.

15. Researcher Problem: $SE_{M_1} = .219$; $SE_{M_2} = .229$; $SE_D = .317$, $t = \dfrac{3.24 - 2.47}{.317} = 2.42$.

    $t_{32, .05} = \pm 2.37$. Reject H_0; the manager should purchase more school apparel with black in it.

Chapter 10

1. $k - 1 = 2$

2. $N - k = 39$

3. 3.24

4. False, the greater heterogeneity between groups and homogeneity within groups, the larger is the calculated value of F.

5. $M_t = \dfrac{(-1.46) + (-.45) + (-1.25)}{3} = -1.05$

   $SS_{between} = (-1.46 - (-1.05))^2 (25) + (-.45 - (-1.05))^2 (25) + (-1.25 - (-1.05))^2 (25)$

   $= 4.20 + 9 + 1 = 14.20$

6. $SS_{new} = s^2 \cdot n - 1 = \left(1.33^2\right)24 = 42.45$

    $SS_{control} = \left(.94^2\right)24 = 21.21$

    $SS_{aspirin} = \left(.53^2\right)24 = 6.74$

    $SS_{within} = 42.45 + 21.21 + 6.74 = 70.40$

7. $MS_{between} = \dfrac{SS_{between}}{k-1} = \dfrac{14.20}{2} = 7.10$

8. $MS_{within} = \dfrac{SS_{within}}{N-k} = \dfrac{70.40}{72} = .98$

9. $F = \dfrac{MS_{between}}{MS_{within}} = \dfrac{7.10}{.98} = 7.24$

10. The critical value for the F distribution with 2 and 72 degrees of freedom and an alpha of .05 is 3.13. 7.24 is located in the alpha area so the decision is made to reject the null hypothesis. A post hoc test of multiple comparisons now needs to be done to determine which alternative hypothesis is the correct one.

11. Research Problem:
Descriptive Information

Group 1: $M = 43.20$, $SS = 22.79$ ($s = 2.39$), $\Sigma X_1 = 216$, $\Sigma X_1^2 = 9354$

Group 2: $M = 37.40$, $SS = 101.20$ ($s = 5.03$), $\Sigma X_2 = 187$, $\Sigma X_2^2 = 7095$

Group 3: $M = 34.80$, $SS = 36.80$ ($s = 3.03$), $\Sigma X_3 = 174$, $\Sigma X_3^2 = 6092$

$M_{total} = 38.47$ $\Sigma X_t = 577$ $\Sigma X_t^2 = 22{,}541$

Using the Computational Approach in the Textbook

$$SS_t = 22{,}541 - \frac{577^2}{15} = 22{,}541 - 22{,}195.27 = 345.73 \text{ (NOTE: } C = 22{,}195.47)$$

$$SS_b = \frac{216^2}{5} + \frac{187^2}{5} + \frac{174^2}{5} - 22{,}195.27 = 9331.20 + 6993.80 + 6055.20 - 22{,}195.27 = 184.93$$

$SS_w = SS_t - SS_b = 345.73 - 184.93 = 160.80$

$df_b = k - 1 = 3 - 1 = 2$ $df_w = N - k = 15 - 3 = 12$

$$V_b = MS_b = \frac{184.93}{2} = 92.465 \qquad\qquad V_w = MS_w = \frac{160.80}{12} = 13.40$$

$$F = \frac{92.465}{13.40} = 6.90 \qquad \text{critical value for } \alpha = .05: 3.88 \qquad \text{critical value for } \alpha = .01: 6.93$$

Using the Deviational Approach

$$SS_b = (43.20 - 38.47)^2(5) + (37.40 - 38.47)^2(5) + (34.80 - 38.47)^2(5) =$$

$$111.8645 + 5.745 + 67.3445 = 184.95$$

$$SS_w = 22.79 + 101.20 + 36.80 = 160.79$$

$$SS_t = SS_b + SS_w = 184.95 + 160.79 = 345.74$$

$$df_b = k - 1 = 3 - 1 = 2 \qquad\qquad df_w = N - k = 15 - 3 = 12$$

$$V_b = MS_b = \frac{184.95}{2} = 92.475 \qquad\qquad V_w = MS_w = \frac{160.79}{12} = 13.40$$

$$F = \frac{92.465}{13.40} = 6.90 \qquad \text{critical value for } \alpha = .05: 3.88 \qquad \text{critical value for } \alpha = .01: 6.93$$

Chapter 11

1. High school GPA and college GPA are positively correlated. As one increases, the other tends to be higher as well.
2. High school GPA is placed on the X axis and college GPA is placed on the Y axis because high GPA is used to predict college GPA.
3. As the correlation becomes stronger, X becomes a better predictor of Y.
4.

X	Y	XY	X^2	Y^2	
1	1	1	1	1	$\Sigma X = 15$
2	2	4	4	4	$\Sigma Y = 15$
3	3	9	9	9	$\Sigma XY = 55$
4	4	16	16	16	$\Sigma X^2 = 55$
5	5	25	25	25	$\Sigma Y^2 = 55$
15	15	55	55	55	

$$SS_X = 55 - \frac{(15)^2}{5} = 55 - 45 = 10$$

$$s_X = \sqrt{\frac{10}{4}} = 1.58$$

$$SS_Y = 55 - \frac{(15)^2}{5} = 55 - 45 = 10$$

111

$$s_Y = \sqrt{\frac{10}{4}} = 1.58$$

$$r = \frac{\dfrac{55 - (3)(3)(5)}{4}}{(1.58)(1.58)} = \frac{2.50}{2.50} = 1$$

5.

X	Y	XY	X^2	Y^2	
1	5	5	1	25	$\Sigma X = 15$
2	4	8	4	16	$\Sigma Y = 15$
3	3	9	9	9	$\Sigma XY = 35$
4	2	8	16	4	$\Sigma X^2 = 55$
5	1	5	25	1	$\Sigma Y^2 = 55$
15	15	35	55	55	

$$SS_X = 55 - \frac{(15)^2}{5} = 55 - 45 = 10$$

$$s_X = \sqrt{\frac{10}{4}} = 1.58$$

$$SS_Y = 55 - \frac{(15)^2}{5} = 55 - 45 = 10$$

$$s_Y = \sqrt{\frac{10}{4}} = 1.58$$

$$r = \frac{\dfrac{35 - (3)(3)(5)}{4}}{(1.58)(1.58)} = \frac{-2.50}{2.50} = -1$$

6. $$b = \frac{rs_Y}{s_X} = \frac{(-1)(1.58)}{1.58} = -1.58$$

$$a = M_Y - bM_X = 3 - (-1)(3) = 3 + 3 = 6$$

$$\hat{Y} = 6 + (-1)(X) = 6 - X$$

7. $SS_Y = 10$

8. $SS_{Y \text{ not explained by } X} = 0$. All five of the $Y - \hat{Y}$ differences are 0.

9. $SS_{Y \text{ explained by } X} = 10$

10. $r^2 = -1^2 = 1$

$$r^2 = \frac{SS_{Y \text{ explained by } X}}{SS_Y} = \frac{10}{10} = 1$$

All the variability (100%) of Y is explained by X. X is a perfect predictor of Y.

11. regression line

12. Researcher Problem: Note that the sum of the cross products is 83,065

Descriptive Statistics

	Mean	Std. Deviation	N
WORKHRS	16.97	15.226	33
GAS$MTH	111.36	85.826	33

Correlations

		WORKHRS	GAS$MTH
WORKHRS	Pearson Correlation	1	.495**
	Sig. (2-tailed)		.003
	Sum of Squares and Cross-products	7418.970	20701.364
	Covariance	231.843	646.918
	N	33	33
GAS$MTH	Pearson Correlation	.495**	1
	Sig. (2-tailed)	.003	
	Sum of Squares and Cross-products	20701.364	235713.636
	Covariance	646.918	7366.051
	N	33	33

**. Correlation is significant at the 0.01 level (2-tailed).

The r is .50 and significant at the .01 level. r^2 is .25. The answers to the two questions are:

1. Yes, there is a relationship between the number of hours worked per week and the amount of money spent on gas.
2. The relationship has moderate strength (.50).

Chapter 12

1. Nominal

2. Goodness of Fit Test and Test of Independence

3. For the Goodness of Fit Test, degrees of freedom equal number of categories - 1. For the Test of Independence, degrees of freedom equal (number of rows - 1) x (number of columns - 1)

4. 9 = 3 x 3

5. 11.07

6. 6.63

7. In the Goodness of Fit test, expected frequencies are computed either based on a) dividing the sample by the number of categories, which is called the frequencies expected by chance, or b) a theoretical prediction.

8. In the Test of Independence, expected frequencies are computed as:

$$\frac{\text{row marginal x column marginal}}{\text{total sample size}}$$

9.

	vote for the candidate	not vote for the candidate
observed frequency (f_o)	185	115
expected frequency (f_e)	195 (65% of the 300 predicted by candidate)	105
$f_o - f_e$	-10	10
$(f_o - f_e)^2$	100	100
$\dfrac{(f_o - f_e)^2}{f_e}$	.51	.95

$$\chi^2 = \Sigma \frac{(f_o - f_e)^2}{f_e} = .51 + .95 = 1.46$$

The critical value for χ^2 with 1 degree of freedom and $\alpha = .05$ is 3.84. The calculated value for χ^2 is smaller than the critical value so we do not reject the null hypothesis.

10.

	Natural Childbirth	Ceasarean Section	
Urban Hospitals	A 40	B 10	50
Rural Hospitals	C 48	D 2	50
	88	12	100

	A	B	C	D
observed frequency (f_o)	40	10	48	2
expected frequency (f_e)	44	6	44	6
$f_o - f_e$	-4	4	4	-4
$(f_o - f_e)^2$	16	16	16	16

114

Central Tendency and Frequency Distribution Assignment

I. For Frequency Distribution 1, combine the women's and men's weights in the class into one data set.

   A. What is the inclusive range?

   B. On your paper, select an appropriate number of intervals, compute the interval size, and construct a frequency distribution table. Include columns for the intervals, frequencies, and relative frequencies.

   C. On the *top half* of the graph paper (see back), graph the data from your frequency table into a frequency polygon. Use a straight edge to make your polygon neat and more accurate.

   D. Compute for this distribution:

      1. mean
      2. median
      3. mode

II. For Frequency Distributions 2 and 3, regard the women and men as two separate distributions, but *graph both frequency distributions on the same graph.*

   A. What are the inclusive ranges for the women's distribution and the men's distribution?

   B. Repeat Step B in Section 1, constructing a frequency distribution table for each gender's weights. Use the intervals for the frequency distribution table developed in Section I. B. again.

   C. On the *bottom half* of the graph paper on the back, plot each gender's data into a *relative frequency* polygon. Make sure to mark the different polygons differently (e.g., red vs. blue line or solid vs. dashed line), use a straight edge to be neat, and include a legend to indicate to the reader which type of line represents which distribution.

   D. Compute for the women:

      1. mean
      2. median
      3. mode

   E. Compute for the men:

      1. mean
      2. median
      3. mode

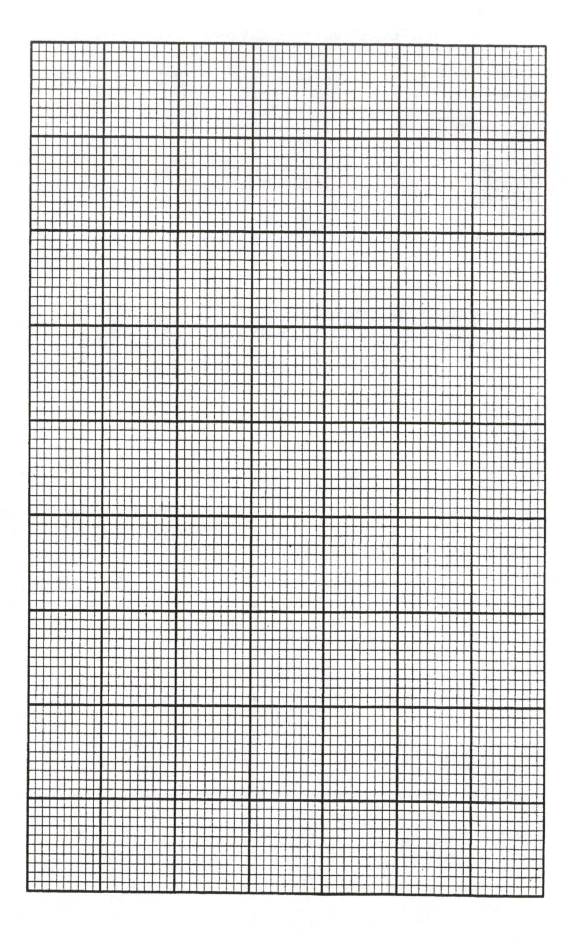

Variability Assignment

For the women's weights and then for the men's weights, calculate the following and indicate your answers below.

1. What is the mean of the women's weights?

2. What is the mean of the men's weights?

3. On a sheet of paper, create a table with the following four column headings on one side of your paper for the women's weights and the same table on the reverse side for the men's weights:

$$X \qquad X^2 \qquad x \qquad x^2$$

4. What is the sum of squares for the women's weights using the deviational formula?

5. What is the sum of squares for the women's weights using the computational formula?

6. What is the sum of squares for the men's weights using the deviational formula?

7. What is the sum of squares for the men's weights using the computational formula?

8. What is the sum of the deviation scores for the women's weights?

9. What is the sum of the deviation scores for the men's weights?

10. What is the variance for the women's weights?

11. What is the standard deviation for the women's weights?

12. What is the variance for the men's weights?

13. What is the standard deviation for the men's weights?

14. Calculate the mean for the overall data set of men's and women's weights using the formula:

$$M_{overall} = \frac{\left[\left(M_{men} \cdot N_{men}\right) + \left(M_{women} \cdot N_{women}\right)\right]}{N_{men} + N_{women}}$$

z Score Assignment

Read each problem carefully. Show all your work. Assume a normal curve and use the normal curve table in the appendix in your textbook to solve the following problems.

1. A person whose raw score converts to a z score of 1.05 standard deviations has what percent of the scores in the distribution below himself/herself?

2. Using the data in #1, what percent of the scores in the distribution are above herself/himself?

3. What is the percent of the total area under the normal curve bordered by −1.39 and +1.39 standard deviations?

4. What is the percent of the total area under the normal curve bordered by −.82 and +.53 standard deviations?

5. If 300 children's test scores are normally distributed, how many of their test scores will fall within 1.00 standard deviation to the right and 1.00 standard deviation to the left of the mean?

For problems 6–10, assume a normal distribution of 3,000 IQ scores with a mean of 100 and a standard deviation of 15.

6. What is the raw score corresponding to a z score of −1.26?

7. What is the z score corresponding to an IQ score of 95?

8. What percent of the population has IQ scores between 105 and 110?

9. How many scores are between 75 and 80?

10. In this distribution, 680 people have test scores that fall to the right and left of the mean between what ±z scores?

Sampling Distribution of the Mean and z Test Assignment

1. What happens to the size of the standard error of the mean if the size of the random samples selected from the population is increased from 50 to 200? Specify the change in size.

2. Assume a sampling distribution of sample means with a mean of 50 and a standard error of the mean of 2.40. What is the probability of selecting a single random sample from the population with a mean smaller than 47?

3. The population of a new LED light bulb has a mean life of 25,000 hours (note: this is the population mean) and population standard deviation of 5000 hours. What is the probability that a sample of 144 light bulbs drawn from this population will have a mean greater than 26,000 hours?

4. The standardized test for nursing content knowledge has a population mean of 110 and a population standard deviation of 12. A class of 64 students takes this test and produces a mean of 104. Using the z test, is this sample of 64 representative of the population? Use α = .05.

5. On the Wechsler Intelligence Scale for Children–Revised, the population mean IQ is 100 and population standard deviation is 15. A class of 25 third grade children is tested with a resulting mean of 94. Does this group have a mean significantly different from the norm group? Use α = .01.

Analysis of Variance (ANOVA) Assignment

The residence halls director wants to determine whether different types of housing make a difference in retaining students in the residence halls. Three different types of residence halls are created:
1. Two bedroom apartments with private bath (APT)
2. Cluster of four individual rooms with communal bathroom (CLU)
3. Floor of two-person rooms with a communal bathroom for the floor (CON).

Seven sophomores were randomly assigned to each residence hall type. At the end of the academic year, the 21 students were asked to indicate on a scale from 1 (terrible) to 10 (wonderful) how they rated their living arrangement. The scores for the seven students for each type of residence hall are:

APT	CLU	CON
7	5	3
8	2	4
8	6	2
5	4	5
9	1	5
4	8	1
8	4	2

1. State H_0 and H_1 (note that there are more than one H_1).

2. Compute ΣX and ΣX^2 for each column.

3. What is ΣX_t? What is ΣX_t^2? What is the overall mean?

4. What is the total sum of squares (SS_t)?

5. What is the sum of squares between groups (SS_{sb})?

6. What is the sum of squares within groups (SS_{sw})?

7. How many degrees of freedom are there for the SS_b and SS_w, respectively?

8. What is the calculated value of F?

9. With $\alpha = .05$, find the critical value for F from the ANOVA table in the back of your book?

10. What do you conclude about H_0?

Here is the Analysis of Variance Table:

SOURCE	SS	df	MS	F
Between groups	54.95	2	27.48	7.24
Within groups	68.29	18	3.79	
Total	123.24			

CPSIA information can be obtained
at www.ICGtesting.com
Printed in the USA
LVHW060258030722
722627LV00001B/5

9 781465 282606